Chapter 1: The Importance of Effecti
- Introduction 9
- History of Risk Management in Construction 9
- Overview of Construction Project Risks 9
- The Role of Effective Risk Management in Construction Projects 10
- The Benefits of Effective Risk Management 10
- The Potential Consequences of Ineffective Risk Management 10
- The Impact of Technology on Risk Management 10
- The Importance of a Collaborative Approach 11
- Conclusion 11

Chapter 2: Effective Risk Identification for Construction Projects 12
- Common Types of Construction Project Risks 12
- Methods for Identifying Risks 13
- Importance of Early Identification 13

Chapter 3: Risk Assessment 15
- Qualitative Risk Assessment 15
- Quantitative Risk Assessment 16
- Tools and Techniques for Risk Assessment 16

Chapter 4: Effective Risk Management Techniques 18
- Techniques for Reducing Risks 18
  - 1. Proper Planning 18
  - 2. Regular Communication 18
  - 3. Continuous Risk Monitoring 18
  - 4. Implementing Safety Practices 19
  - 5. Incorporating Quality Control Measures 19
- Risk Transfer and Insurance 19
  - 1. Contractual Risk Transfer 19
  - 2. Obtaining Construction Insurance 20
  - 3. Risk Pooling 20
- Risk Avoidance 20
  - 1. Thorough Screening of Contractors 20
  - 2. Conducting Due Diligence on Suppliers 20
  - 3. Avoiding High-Risk Projects 21

Chapter 5: Developing an Effective Risk Response Plan for Construction Projects .. 22

- Developing a Risk Response Plan 22
  - Avoid 22
  - Transfer 22
  - Mitigate 22
  - Accept 23
- Contingency Planning 23
- Business Continuity Planning 23
- Conclusion 23

Chapter 6: Monitoring and Control of Risks 25
- Importance of Regularly Monitoring Risks 25
- Tracking Risk Responses 25
- Handling Unforeseen Risks 26
- In Summary 26

Chapter 7: Communication for Effective Risk Management 27
- Communication Strategies for Risk Management 27
- Stakeholder Involvement 28
- Transparency in Communication 28

Chapter 8: Integrating Risk Management into Project Planning 30
- Integrating Risk Management into Project Planning 30
- Impact of Risk Management on Project Schedule and Budget 30
- Risk Management Best Practices 31

Chapter 9: Effective Risk Management Techniques for Construction Projects - Managing Project Budgeting and Cost Control 33
- Estimating and Allocating Contingency Funds 33
- Managing Cost Overruns 34
- Value Engineering 34
- Conclusion 34

Chapter 10: Contracts and Risk Management 36
- Importance of Clear Contracting 36
- Legal Considerations for Risk Management 36
- Mitigating Risks Through Contracts 37
- Conclusion 37

Chapter 11: Safety and Risk Management 39
- Relationship Between Safety and Risk Management 39
- Identifying and Mitigating Safety Risks 39
- Ensuring Compliance with Safety Regulations 40

Chapter 12: Human Resource Management and Risk 41
Impact of Human Factors on Risk 41
Training and Development for Risk Management 41
Managing Human Error 42

Chapter 13: Quality Management and Risk 43
Quality Control Measures to Minimize Risks 43
Risk of Defects and Delays 43
Ensuring Quality in Construction Projects 44

Chapter 14: Identifying and Managing Environmental Risks 45
Importance of Adhering to Environmental Regulations 45
Environmental Risk Management Process 45
Use of Sustainable Practices 46
Benefits of Proper Environmental Risk Management 46
Involving Stakeholders in Environmental Risk Management 46
Conclusion 46

Chapter 15: Understanding and Managing supply chain risks 48
Understanding Supply Chain Risks 48
Managing Risks in the Supply Chain 48
Building Resilient Supply Chain Partnerships 49

Chapter 16: Technology and Risk Management 51
Role of Technology in Risk Management 51
Tools and Software for Risk Management 51
Integrating Technology into Risk Management Plans 52

Chapter 17: Case Studies of Successful Risk Management in Construction Projects 53
Case Study 1: The Burj Khalifa Project 53
Case Study 2: The Olympic Stadium in London 53
Common Mistakes and How to Avoid Them 54
    Mistake #1: Not Conducting Proper Risk Assessments 54
    Mistake #2: Failure to Update Risk Management Strategies 54
    Mistake #3: Lack of Communication 54
    Mistake #4: Ignoring Lessons Learned 55
    Mistake #5: Not Involving Experts 55

Chapter 18: International Perspectives on Risk Management 56
Cultural Differences in Risk Management 56
Best Practices in International Risk Management 57

Chapter 19: Legal and Ethical Considerations ... 59
    Legal Obligations for Risk Management ... 59
    Ethical Principles in Managing Project Risks ... 59
    Dealing with Conflict of Interest ... 60
    Closing Thoughts ... 60
Chapter 20: Managing Crisis Risks in Construction Projects ... 61
    Understanding and Preparing for Crisis Situations ... 61
    Crisis Communication ... 61
    Strategies for Managing Crisis Risks ... 62
Chapter 21: The Importance of Design in Risk Management ... 63
    Role of Design in Risk Management ... 63
    Identifying and Mitigating Design Risks ... 63
    Designing for Resilience ... 64
    Conclusion ... 64
Chapter 22: Effective Risk Management in Different Project Delivery Methods ... 65
    Comparison of Different Delivery Methods ... 65
    Choosing the Right Method to Minimize Risks ... 65
    Case Studies of Successful Projects Using Various Delivery Methods ... 66
    In Conclusion ... 67
Chapter 23: Managing Risks in Mega-Projects ... 68
    Unique Risks in Mega-Projects ... 68
    Challenges and Solutions in Managing Risks in Large-Scale Projects ... 69
Chapter 24: Identifying and Managing Risks in Public-Private Partnership Projects ... 71
    Understanding Public-Private Partnerships ... 71
    Risk Identification in PPP Projects ... 71
    Managing Risks in PPP Projects ... 72
        1. Establish a Clear Allocation of Risks ... 72
        2. Conduct Extensive Due Diligence ... 72
        3. Define a Communication Plan ... 72
        4. Include Risk Sharing Mechanisms ... 72
    Case Studies of Successful Partnerships ... 73
        1. Indiana Toll Road ... 73
        2. London Underground ... 73
    Conclusion ... 73
Chapter 25: Effective Communication with Project Teams ... 74

Strategies for Effective Communication 74
Importance of Team Collaboration in Risk Management 75
Managing Diverse Project Teams 75
Conclusion 76

Chapter 26: Stakeholder Engagement and Relationship Building for Effective Risk Management in Construction Projects 77
Stakeholder Analysis and Engagement 77
Mitigating Reputational Risks 78
Building Strong Stakeholder Relationships 78

Chapter 27: Identifying and Managing Risks in Sustainable Construction 80
Integrating Sustainability into Risk Management Plans 80
Identifying Risks in Sustainable Construction 80
Managing Risks in Sustainable Construction 81
Incorporating Sustainability into Risk Response Planning 82
In Conclusion 82

Chapter 28: Identifying and Managing Risks in the Maintenance Phase 83
Identifying Risks in the Maintenance Phase 83
Managing Risks in the Maintenance Phase 83
Preventative Maintenance Strategies 84
Case Studies of Successful Maintenance Risk Management 85

Chapter 29: Effective Risk Management for Small and Medium-Sized Construction Companies 87
Unique Challenges for Small and Medium-Sized Companies 87
   1. Limited Financial Resources 87
   2. Lack of Expertise 87
   3. Limited Access to Information 87
Strategies for Minimizing Risks with Limited Resources 88
   1. Go back to Basics 88
   2. Develop a Risk Management Plan 88
   3. Prioritize Risks 88
   4. Invest in Training and Development 88
   5. Leverage Technology 89
   6. Collaborate with Larger Companies 89
   7. Continuously Evaluate and Improve 89

Chapter 30: Project Portfolio Risk Management 90
Project Portfolio Risk Management 90

- Prioritizing Risks Across Multiple Projects 90
- Sharing Lessons Learned Across Projects 91

Chapter 31: Managing Risks in the Face of Change 92
- Impact of Change on Project Risks 92
- Strategies for Managing Unexpected Changes 92
- Communicating Changes to Stakeholders 93

Chapter 32: Leveraging Data Analytics for Effective Risk Management 95
- Using Data to Identify and Manage Risks 95
- Predictive Analytics for Risk Mitigation 95
- Integrating Data Analysis into Risk Management Processes 96

Chapter 33: Managing Difficult Clients and Subcontractors 97
- Tools and Techniques for Dealing with Difficult Clients and Subcontractors 97
- Maintaining Professional Relationships 97
- Negotiating Contracts and Agreements 98
- Conclusion 98

Chapter 34: Managing Conflicting Priorities in Risk Management 100
- Balancing Project Objectives with Risk Management 100
- Prioritizing Risks Based on their Impact and Likelihood 100
- In Conclusion 101

Chapter 35: Risk Management in Public Sector Projects 103
- Unique Challenges in Managing Risks in Public Projects 103
- Strategies for Effective Risk Management in the Public Sector 103

Chapter 36: Managing Risks in International Construction Projects 106
- Identifying and Mitigating Risks in International Projects 106
- Cultural Differences and Legal Considerations 106
- Ensuring Compliance with Local Regulations 107

Chapter 37: Understanding Financial Risks in Construction 108
- Strategies for Managing Cost-Related Risks 108
- Financial Reporting and Budgeting for Risk Management 108
- Conclusion 108

Chapter 38: Legal Disputes and Construction Projects 110
- Common Legal Disputes in Construction Projects 110
- Strategies for Avoiding and Resolving Disputes 111
- Working with Legal Counsel 111

Chapter 39: Risk Management and Continuous Improvement 113

    Incorporating Risk Management into Continuous Improvement Processes........... 113
    Using Lessons Learned to Improve Future Projects.. 113
Chapter 40: Risk Management for Infrastructure Projects... 115
    Unique Challenges and Risks in Infrastructure Projects.. 115
        Inadequate Planning.. 115
        Environmental and Regulatory Compliance.. 115
        Weather and Natural Disasters.. 116
        Political and Economic Risks... 116
        Resource Availability.. 116
    Strategies for Managing Risks in Building and Transport Projects.................... 117
        Risk Assessment and Identification... 117
        Contingency Planning... 117
        Effective Communication and Stakeholder Engagement.............................. 117
        Continuous Monitoring and Control... 117
    In Conclusion... 117
Chapter 41: Role of Technology in Streamlining Risk Management Processes...... 119
    Integration with Building Information Modeling (BIM).. 119
    Virtual and Augmented Reality for Risk Management.. 119
    Conclusion... 119
Chapter 42: Understanding Political and Economic Risks in Construction Projects....... 121
    Understanding Political and Economic Factors that Affect Construction Projects....... 121
    Strategies for Mitigating Risks due to Changes in Government or Economy........ 121
Chapter 43: Risk Management during Construction Delays..................................... 124
    Types and Causes of Construction Delays... 124
        Weather-Related Delays... 124
        Design and Planning Delays... 124
        Material and Equipment Delays... 124
        Labor Shortages... 124
        Unforeseen Site Conditions.. 125
    Causes of Construction Delays.. 125
        Poor Communication.. 125
        Inadequate Planning and Risk Management... 125
        Poor Project Management.. 125
        Scope Changes.. 125

- External Factors 126
- Risk Management Strategies to Minimize Delay Risks 126
  - Identify and Assess Risks 126
  - Develop Contingency Plans 126
  - Regularly Monitor and Control Risks 127
  - Maintain Good Communication 127
  - Utilize Technology 127
- Communicating with Stakeholders during Delays 127
  - Be Transparent and Honest 127
  - Explain the Impact of the Delay 128
  - Discuss Potential Solutions 128
  - Document Communications 128

Chapter 44: Identifying and Managing Natural and Man-Made Disaster Risks 129
- Understanding the Risks 129
- Strategies for Building Resilient Structures and Processes 129
  - Design and Construction Techniques 129
  - Emergency Response Plan 129
  - Backup and Recovery Systems 130
  - Insurance and Legal Protection 130
- Cultivating a Culture of Disaster Resilience 131
- Conclusion 131

Chapter 45: Understanding Cybersecurity Risks in Construction Projects 132
- Data Protection and Security Measures 132
- Strategies for Managing Cybersecurity Risks 133
- Understanding Cybersecurity Risks in Construction Projects 133
- In Conclusion 134

Chapter 46: Identifying and Managing Risks in BIM Projects 135
- Identifying and Managing Risks in BIM Projects 135
- Collaborative Risk Management in BIM 135
- Integrating BIM into Risk Management Processes 135
- Conclusion 135

Chapter 47: Managing Risks in Design-Build Projects 137
- Unique Risks in Design-Build Projects 137
- Strategies for Managing Risks in Integrated Project Delivery 137
- Case Studies of Successful Design-Build Projects 138

In Conclusion	138
Chapter 48: Risks Management in Public-Private Infrastructure Partnerships	**140**
Unique Risks in P3 Projects	140
Political Risk	140
Financial Risk	140
Legal Risk	140
Social Risk	141
Collaborative Risk Management with Public and Private Partners	141
Early Risk Identification and Allocation	141
Risk Allocation through Contracts	141
Regular Communication and Collaboration	142
Shared Risk Management Tools and Techniques	142
Case Studies of Successful P3 Projects	142
Denver International Airport Expansion Project, USA	142
Kempenfelt Water Treatment Plant, Canada	142
N1/N2 Ring Roads, Ireland	143
Chapter 49: Risk Management for Green Building Projects	**144**
Identifying and Managing Risks in Sustainable Construction	144
Strategies for Minimizing Risks in Green Building Projects	144
Certification and Verification Risks	145
In Conclusion	145

Chapter 1: The Importance of Effective Risk Management in Construction Projects

Introduction

Effective risk management is crucial in ensuring the success of construction projects. With the increasing complexity and magnitude of construction projects, the potential risks involved have also multiplied. Failure to properly manage these risks can result in project delays, cost overruns, safety hazards, and even litigation. Therefore, it is imperative for construction companies to have a solid risk management plan in place to minimize the negative impacts of unforeseen events.

History of Risk Management in Construction

Risk management has always been a part of the construction industry, in one form or another. However, it was not until the 1960s and 1970s that formal techniques and methodologies were developed to address risks in construction projects. With the introduction of project management as a discipline, there was a growing recognition of the need to identify, assess, and manage risks in construction projects. This gave rise to various risk management frameworks, such as the Monte Carlo simulation and the risk matrix approach.

Overview of Construction Project Risks

Construction projects are highly vulnerable to various risks, both internal and external. These risks can be categorized into four main types: financial, physical, legal, and environmental. Financial risks include budget overruns, cash flow issues, and market fluctuations. Physical risks involve safety hazards, material shortages, and weather conditions. Legal risks can arise from contractual disputes or non-compliance with regulations. Environmental risks pertain to the impact of the project on the surrounding ecosystem. It is essential for project managers to have a comprehensive understanding of these risks and how they may influence the project outcome.

The Role of Effective Risk Management in Construction Projects

Effective risk management can help construction companies in various aspects, including cost control, schedule management, and reputation protection. By identifying potential risks and implementing strategies to mitigate or eliminate them, costs can be reduced, and project schedules can be maintained. Furthermore, proper risk management can also enhance the company's reputation by demonstrating a commitment to responsible project execution and stakeholder satisfaction.

The Benefits of Effective Risk Management

Implementing effective risk management techniques can bring numerous benefits to construction projects. These include improved project performance, increased stakeholder satisfaction, and better decision-making. By proactively managing risks, project teams can stay on top of potential issues and handle them efficiently, leading to higher quality project outcomes. Stakeholder satisfaction can also be improved by addressing any concerns or risks that may affect them. Finally, effective risk management can also aid in better decision-making by providing accurate and reliable information on potential risks and their impact on the project.

The Potential Consequences of Ineffective Risk Management

On the other hand, failing to properly manage risks can result in significant negative consequences for construction projects. Delays, cost overruns, and rework due to unanticipated risks can all impact project success and damage company reputation. In the worst-case scenario, poor risk management can lead to legal disputes, safety hazards, and even project failure.

The Impact of Technology on Risk Management

The advancement of technology has enabled construction companies to adopt more sophisticated risk management techniques. With the help of various software and tools, project teams can identify and analyze risks more efficiently, accurately predict project outcomes, and monitor risks in real-time. The use of building information modeling (BIM) technology has also allowed for greater collaboration and risk management among project stakeholders.

The Importance of a Collaborative Approach

While project managers are ultimately responsible for risk management, it is crucial to involve all stakeholders in the process. A collaborative approach enables different perspectives and expertise to be taken into consideration, leading to more thorough risk identification and mitigation plans. Involving stakeholders also fosters a sense of ownership and accountability, leading to a more effective risk management process.

Conclusion

In conclusion, effective risk management is essential for the success of construction projects. The history of risk management in construction, an overview of project risks, and the potential consequences of poor risk management all highlight the need for a robust risk management plan. By proactively identifying and managing risks, construction companies can ensure project success, maintain stakeholder satisfaction, and protect their reputation. With the increasing complexity and risks involved in construction projects, it is more important than ever to prioritize and invest in effective risk management techniques.

Chapter 2: Effective Risk Identification for Construction Projects

Construction projects are complex endeavors that involve numerous stakeholders, various scopes of work, and a multitude of moving parts. With such complexity, comes inherent risks that can have a significant impact on the project's success. As such, it is crucial for construction companies to have effective risk management techniques in place to identify and mitigate potential risks. In this chapter, we will explore the common types of construction project risks, methods for identifying risks, and the importance of early identification.

Common Types of Construction Project Risks

Before diving into methods for identifying risks, it is essential to understand the common types of risks that are typically associated with construction projects. These risks can range from financial and schedule-related risks to safety and environmental risks. Some of the most common risks include:

- Budget and cost overruns

- Delays in construction schedule

- Safety incidents and accidents

- Changes in project scope

- Inadequate planning and design

- Inflation and economic risks

- Disputes and conflicts with stakeholders

- Weather and natural disasters

- Inaccurate estimations of resources and materials

- Changes in regulations and permits

Each type of risk can have a significant impact on the project's budget, schedule, and overall success. Therefore, it is vital to have a systematic approach to identify and mitigate these risks throughout the project's lifecycle.

Methods for Identifying Risks

Identifying risks is the first step in effective risk management. It involves a systematic process of identifying potential risks, analyzing their likelihood and impact, and developing strategies to mitigate or avoid them. There are various methods that construction companies can use to identify risks, including but not limited to:

- Brainstorming: This involves gathering a group of project stakeholders to identify potential risks based on their experience and expertise.

- Checklist approach: Using a risk checklist to identify common risks associated with construction projects.

- Historical data analysis: Examining past projects and their associated risks to identify potential risks for the current project.

- Interviews: Conducting interviews with project team members and other stakeholders to understand their perspective on potential risks.

- SWOT analysis: Analyzing the strengths, weaknesses, opportunities, and threats of the project to identify internal and external risks.It is crucial to note that no single method is sufficient to identify all potential risks. Therefore, a combination of these methods can be used to ensure a comprehensive risk identification process.

Importance of Early Identification

The success of a construction project heavily relies on identifying potential risks as early as possible. Early identification allows project teams to develop a robust risk response plan and implement it in a timely manner. It can also help avoid delays and additional costs associated with addressing risks later in the project's lifecycle.

Additionally, early identification can lead to a more proactive approach to risk management, allowing project teams to take preventative measures to minimize the impact of potential risks. Moreover, early identification of risks enables project teams to evaluate the potential impact on the project budget, schedule, and quality. This allows for better risk prioritization, ensuring that the most critical risks are addressed first. Not only does this save time and resources, but it also allows for better allocation of resources towards risk mitigation strategies. Another benefit of early risk identification is increased stakeholder confidence. When project teams proactively identify and address potential risks, stakeholders can trust that the project is being managed effectively. This can lead to better relationships with stakeholders, and in turn, contribute to project success.

In conclusion, effective risk identification is a crucial step in risk management for construction projects. With the right methods and a proactive approach, project teams can identify and mitigate potential risks, leading to a successful project outcome. In the next chapter, we will explore the next steps in effective risk management, including risk assessment and mitigation.

Chapter 3: Risk Assessment

Qualitative Risk Assessment

Risk assessment is a crucial step in effective risk management for construction projects. It involves identifying potential risks and analyzing their impact on the project. However, the risk assessment process can be complex and challenging, as it requires the consideration of multiple factors and uncertainties. In order to simplify this process, qualitative risk assessment techniques are often used. These techniques allow project managers to assess risks based on their likelihood and impact, without the need for detailed and quantitative data. One of the key advantages of qualitative risk assessment is its simplicity.

Unlike quantitative risk assessment, which requires extensive data and calculations, qualitative risk assessment relies on expert judgment and subjective evaluation. This makes it more accessible and easier to understand for project stakeholders who may not have a technical background. It also allows for a quicker assessment of risks, which is particularly beneficial in time-sensitive construction projects where rapid decision-making is crucial. Another advantage of qualitative risk assessment is its flexibility. This approach allows for a wide range of risks to be considered, even those that may be difficult to measure or quantify. For construction projects, where risks can come from various sources such as design changes, adverse weather conditions, or human error, this flexibility is essential. It also allows for risks to be assessed at different stages of the project, enabling project managers to take appropriate actions to mitigate risks as they arise.

Conducting a qualitative risk assessment involves using techniques such as risk probability and impact assessment, risk categorization, risk ranking, and risk appetite analysis. Risk probability and impact assessment involves identifying and evaluating risks based on their likelihood and potential consequences. Risk categorization involves grouping risks into categories, such as safety, financial, environmental, and schedule risks, to create a better understanding of the different types of risks that may impact the project. Risk ranking involves prioritizing risks based on their level of severity, with the most significant risks receiving the highest priority. Lastly, risk appetite analysis allows project stakeholders to determine their tolerance for risk and set risk management strategies accordingly.

Quantitative Risk Assessment

While qualitative risk assessment can provide a useful overview of potential risks, there are limitations to this approach. As qualitative risk analysis relies on expert judgment and subjective evaluation, it can be prone to bias and may not provide a comprehensive understanding of risk likelihood and impact. This is where quantitative risk assessment comes in. This approach involves using numerical data and statistical analysis to assess risks, providing a more precise and quantifiable measure of risk probability and impact. Quantitative risk assessment techniques use mathematical models and tools to analyze the potential impact of identified risks. These tools can range from simple spreadsheets to complex software programs, such as Monte Carlo simulation. The Monte Carlo simulation model works by running multiple simulations of a project, considering various risk factors and their potential impacts. This allows project managers to identify the most probable risks and their potential consequences and to develop mitigation strategies accordingly. The use of quantitative risk assessment can provide a more comprehensive and accurate understanding of risks and their potential impact. It also allows for a more detailed analysis of risks and their interdependencies, enabling project managers to prioritize and manage risks more effectively. Additionally, quantitative risk assessment techniques can provide a foundation for cost and schedule contingency planning, helping project managers to create a more accurate project budget and timeline. However, the use of quantitative risk assessment techniques requires a significant amount of data and expertise. It also comes at a higher cost than qualitative risk assessment. Therefore, for smaller projects with limited resources, qualitative risk assessment techniques may be more practical and suitable.

Tools and Techniques for Risk Assessment

With the advancements in technology, there are now numerous tools and techniques available to assist with risk assessment for construction projects. These tools range from traditional spreadsheets and checklists to more advanced software programs that utilize data and simulations to analyze risks. Some of the commonly used tools and techniques for risk assessment include risk breakdown structures, risk matrices, risk registers, and risk maps. A risk breakdown structure (RBS) is a hierarchical representation of risks that breaks them down into categories and subcategories, making it easier to identify and manage risks. It also allows project managers to track the progress of risk management efforts and monitor the status of risks. A risk matrix is a visual tool that helps to assess risks based on their likelihood and impact, usually on a scale of low to high. This allows project managers to prioritize risks based on their severity and develop appropriate mitigation strategies. A risk register is a document that lists all identified risks, their potential impact, and the actions taken to manage them. It is a useful tool for monitoring risks and their status throughout the project. Risk maps are graphical representations of risks, often in the form of a heat map, that helps to visualize the probability and severity of risks at different stages of the project. This can provide project stakeholders with a quick and clear understanding of potential risks and aid in decision-making for risk management. In addition to these tools, there are also various software programs available that can assist with risk assessment, such as @RISK, Palisade DecisionTools Suite, and Primavera Risk Analysis. These programs utilize data and simulations to provide a more comprehensive and accurate assessment of risks, and can greatly aid in risk management for construction projects.

In conclusion, risk assessment is a critical step in effective risk management for construction projects. While qualitative risk assessment techniques provide a quick and accessible way to identify and assess risks, quantitative risk assessment techniques offer a more detailed and precise analysis of risks and their potential impact. By utilizing the right tools and techniques for risk assessment, project managers can improve decision-making, prioritize risks, and develop effective risk management strategies to ensure the successful delivery of construction projects.

Chapter 4: Effective Risk Management Techniques

Techniques for Reducing Risks

Effective risk management involves implementing strategies to reduce the likelihood and impact of potential risks. Here are some techniques that can help construction project teams minimize risks:

1. Proper Planning

The most effective way to reduce risks in construction projects is through proper planning. This includes conducting a thorough risk assessment and identifying potential risks that could arise during the project. By anticipating potential issues, teams can create contingency plans and allocate resources accordingly to mitigate those risks.

2. Regular Communication

Clear and regular communication between project stakeholders is crucial in reducing risks. This involves keeping everyone involved informed about any changes or developments in the project and addressing concerns and issues promptly. By maintaining open lines of communication, project teams can address potential risks before they escalate and cause significant problems.

3. Continuous Risk Monitoring

With large and complex construction projects, risks can arise at any stage. Therefore, it is essential to monitor and reassess risks throughout the project's duration. This allows project teams to stay updated on changing conditions and adjust risk management strategies accordingly.

4. Implementing Safety Practices

Construction sites are inherently hazardous environments, and accidents can occur if proper safety practices are not in place. By implementing safety protocols, providing necessary training, and enforcing these practices, construction project teams can reduce the likelihood of accidents and minimize project interruptions.

5. Incorporating Quality Control Measures

Quality control measures are crucial for ensuring the construction project meets the required standards and specifications. By continuously monitoring quality throughout the project, teams can identify any potential risks that could compromise the project's success and take corrective action.

Risk Transfer and Insurance

While risks can be reduced through proactive measures, some risks cannot be completely eliminated. In such cases, it is essential to have a risk transfer and insurance plan in place. Here are some strategies for transferring risks and obtaining insurance for construction projects:

1. Contractual Risk Transfer

Contractual risk transfer involves shifting the responsibility of potential risks to another party through legally binding agreements. For example, a construction contract may specify that the contractor is responsible for any accidents or damages that occur during the project. In such cases, the contractor would be required to have insurance coverage to mitigate any potential risks.

2. Obtaining Construction Insurance

Construction insurance provides coverage for potential risks that could arise during the project, such as property damage, accidents, and injuries. Contractors and project owners can obtain different types of insurance, such as builder's risk insurance, general liability insurance, and workers' compensation insurance. Having this type of coverage can protect the project from suffering extensive financial losses in the event of any unforeseen circumstances.

3. Risk Pooling

Risk pooling involves spreading the risk across multiple parties to minimize its impact. For example, multiple contractors working on a project could come together to form a risk pool and contribute to a collective insurance fund. If any risks occur, the financial burden would be shared among all the contractors, reducing the impact on individual parties.

Risk Avoidance

Risk avoidance involves taking proactive measures to completely avoid potential risks. While this may not always be feasible, certain steps can be taken to minimize the likelihood of risks occurring.

1. Thorough Screening of Contractors

One way to avoid potential risks is by thoroughly screening contractors before hiring them for the project. This could include checking their credentials, experience, and track record to ensure they are qualified and reliable.

2. Conducting Due Diligence on Suppliers

Suppliers play a crucial role in the success of a construction project, and selecting the wrong supplier could lead to significant risks. By conducting due diligence on suppliers and ensuring they have a good reputation and can provide high-quality materials and services, project teams can avoid potential risks and delays.

3. Avoiding High-Risk Projects

Some projects may present significant risks that may not be worth taking on. It is essential to carefully evaluate the scope, complexity, and potential challenges of a project before deciding to undertake it. Avoiding high-risk projects can prevent unwanted complications and protect the project team from potential financial losses.

In conclusion, effective risk management involves a combination of techniques, including proper planning, communication, and continuous monitoring. Additionally, having a risk transfer and insurance plan in place and taking steps to avoid potential risks can mitigate any impact on the project's success. By incorporating these strategies, construction project teams can effectively manage risks and ensure the project's success.

Chapter 5: Developing an Effective Risk Response Plan for Construction Projects

In the previous chapters, we have discussed the importance of identifying and assessing risks in construction projects. However, risk management does not stop there. The next crucial step is developing a risk response plan to address these potential risks. In this chapter, we will delve into the key elements of an effective risk response plan, including contingency planning and business continuity planning.

Developing a Risk Response Plan

A risk response plan is a proactive approach to addressing potential risks before they occur, minimizing their impact on the project's progress and budget. Before developing a response plan, it is essential to understand the types of risks that could potentially occur and their likelihood of happening. This information can be obtained from the risk assessment conducted in the earlier stages of the project.Once these risks have been identified and assessed, the project team can then develop appropriate response strategies. These strategies can fall into four categories: avoid, transfer, mitigate, and accept.

Avoid

One way to respond to risks is to avoid them altogether. This can be achieved by altering the project plan or approach to eliminate the possibility of the risk occurring. For example, if a project's schedule is at risk due to a delay in material delivery, the project team can switch to a different supplier or adjust the project timeline to account for the delay.

Transfer

Another response strategy is to transfer the risk to another party, typically through insurance or contractual agreements. Transferring risk can reduce the financial burden on the project in case of an adverse event. It is essential to carefully review and assess contracts and insurance policies to ensure all potential risks are covered.

Mitigate

Risk mitigation involves taking proactive measures to reduce the likelihood or impact of a potential risk. This can include implementing safety protocols, using quality materials, or conducting regular inspections. By mitigating risks, the project team can significantly reduce their potential impact on the project.

Accept

Not all risks can be avoided, transferred, or mitigated. In such cases, the best strategy is to accept the risk and prepare contingency plans to address any potential consequences. Accepting risks means acknowledging their existence and having plans in place to handle them if they occur.

Contingency Planning

Contingency planning involves developing alternative strategies and action plans to address potential risks and minimize their impact on the project. These plans serve as a safety net for the project, providing a roadmap to follow if a risk occurs. To develop an effective contingency plan, the project team must work closely with all stakeholders to identify potential risks and their likelihood of occurring. This information will help determine the appropriate level of contingency to allocate in the project budget.Contingency plans should also include clear protocols and procedures for handling risks, as well as designated individuals responsible for implementing them. It is crucial to regularly review and update contingency plans as the project progresses and new risks emerge.

Business Continuity Planning

In addition to contingency planning, business continuity planning is also crucial for risk management in construction projects. Business continuity planning involves developing procedures and protocols to ensure that the project can continue to operate in the event of a risk occurrence. This plan should include strategies for communicating with stakeholders, moving resources to alternative locations, and re-establishing critical project functions. With a well-developed business continuity plan, the project team can minimize the impact of a risk event on the project's timeline and budget.

Conclusion

Effective risk management in construction projects requires a comprehensive and proactive approach. Developing a risk response plan that includes contingency planning and business continuity planning is crucial to minimize the impact of potential risks and keep the project on track. By carefully assessing and addressing risks before they occur, project teams can ensure the success of their construction projects.

Chapter 6: Monitoring and Control of Risks

Effective risk management is an essential component of any successful construction project. However, it is not enough to simply identify and assess risks at the beginning of a project. Risks are dynamic and can change over time, which is why regular monitoring and control are crucial to the success of a project. In this chapter, we will discuss the importance of regularly monitoring risks, tracking risk responses, and handling unforeseen risks.

Importance of Regularly Monitoring Risks

Risk monitoring is the process of continuously reviewing and assessing identified risks throughout the duration of a project. This allows for early detection of any changes in the risk landscape and allows for timely response and mitigation measures to be implemented. Regular monitoring also ensures that risks do not escalate into larger issues that could threaten the project's success. One of the main benefits of regularly monitoring risks is that it helps to keep the project on track. By proactively identifying and addressing risks, project managers can prevent delays, cost overruns, and other issues that could impact the project's timeline and budget. This is especially important in the construction industry where delays and unexpected costs can have a significant impact on the project's bottom line.Another important aspect of risk monitoring is the ability to identify emerging risks. Risks can evolve and new risks can arise throughout the course of a project, and without regular monitoring, they may go unnoticed until it's too late. By keeping a close eye on potential risks, project managers can address them before they become major problems, saving both time and money.

Tracking Risk Responses

After risks have been identified and assessed, a risk response plan should be developed. This plan outlines the measures that will be taken to mitigate, transfer, or accept risks. But the work doesn't stop there. It is essential to track the progress of risk responses to ensure that they are effective and to make adjustments as needed. Tracking risk responses is important because it allows project managers to evaluate the success of their risk management strategies. If a risk response is not effective, it's essential to identify the root cause and find an alternative solution. This ongoing evaluation and adjustment help to ensure that risks are properly addressed and do not have a negative impact on the project. Another benefit of tracking risk responses is that it allows project managers to allocate resources more efficiently. By understanding which risks have been successfully addressed and which are still outstanding, project managers can prioritize their efforts and allocate resources where they are most needed. This can also help to prevent unnecessary expenditures on low-risk items, saving the project money in the long run.

Handling Unforeseen Risks

No matter how well a project is planned, there is always the potential for unforeseen risks to arise. These risks are often unpredictable and can have a significant impact on the project if not addressed promptly. That's why it's crucial to have a plan in place for handling unforeseen risks. One effective way to deal with unforeseen risks is by establishing a contingency plan. This is a predetermined course of action to address unexpected risks and keep the project on track. A contingency plan should also include a contingency budget that can be used to address unforeseen risks without disrupting the main project budget. Another key aspect of handling unforeseen risks is communication. Project managers must ensure that all stakeholders are kept informed of any new risks and the measures being taken to address them. This helps to manage expectations and maintain transparency throughout the project, fostering trust and minimizing potential conflicts.

In Summary

Regularly monitoring and controlling risks is a crucial part of effective risk management for construction projects. It helps to keep the project on track, identify emerging risks, evaluate risk response effectiveness, and allocate resources more efficiently. Additionally, having a plan in place for handling unforeseen risks is essential to mitigate their impact on the project's success. By following these practices, project managers can effectively manage risks and increase the chances of a project's success.

Chapter 7: Communication for Effective Risk Management

Effective communication is a crucial aspect of risk management in construction projects. It involves the transmission and exchange of information between project stakeholders to ensure everyone is on the same page. In this chapter, we will explore various communication strategies, the role of stakeholders, and the importance of transparency in communication for effective risk management in construction projects.

Communication Strategies for Risk Management

The success of any construction project depends heavily on communication. It is the key to identifying, assessing, and mitigating risks. With effective communication strategies in place, project managers can ensure that all team members are informed about potential risks and their corresponding response plans. Here are some communication strategies that can help in risk management:

1. Clear and concise messaging: In a complex and constantly changing construction environment, it is essential to keep all communication clear and concise. This ensures that all stakeholders understand the risks and their impact on the project.

2. Regular communication: It is important to establish a system of regular communication to keep all stakeholders informed about project progress and any changes in risk management strategies.

3. Use of multiple channels: With various communication channels available, it is vital to use a combination of them to ensure that all stakeholders receive the necessary information. This can include emails, project management software, progress reports, and team meetings.

4. Listen and be open to feedback: Communication is a two-way process, and it is essential to listen to the concerns and suggestions of project stakeholders, including team members, clients, and subcontractors. This can provide valuable

insights into potential risks and help improve risk management strategies.

5. Identify a communication lead: Designating a communication lead can help streamline the flow of information between stakeholders and ensure that all communication is consistent and accurate.

Stakeholder Involvement

Risk management is a collaborative effort that involves all project stakeholders. It is crucial to involve stakeholders in the risk management process to ensure shared responsibility and accountability. Here are some ways to involve stakeholders in effective risk management:

1. Project kickoff meetings: The initial project kickoff meeting is an excellent opportunity to involve stakeholders and discuss potential risks and risk management strategies.

2. Team meetings: Regular team meetings provide an opportunity to discuss and address any new or existing risks and discuss ways to mitigate them.

3. Client meetings: It is crucial to involve clients in risk management discussions as they have a vested interest in the project's success. Regular client meetings can help identify potential risks and agree on the best risk management approach.

4. Subcontractor involvement: Subcontractors play a critical role in construction projects, and it is important to involve them in risk management discussions. They can provide valuable input on potential risks and suggest ways to mitigate them.

5. Include risk management in team responsibilities: When every team member understands their role and responsibilities in risk management, it can improve the effectiveness of risk management strategies.

Transparency in Communication

Transparency is key in any successful project, and it is especially important in risk management. It involves openness, honesty, and providing all stakeholders with the necessary information to make informed decisions. Transparency in communication can improve trust and collaboration among stakeholders and ensure that everyone is working towards the same goals.

In risk management, transparency can involve:

- Sharing risk assessment reports with all stakeholders
- Discussing potential risks with clients and subcontractors
- Providing regular updates on risk management strategies and any changes made
- Addressing concerns and providing explanations for risk management decisions
- Encouraging open and honest communication among all project stakeholders

Transparency in communication can also help in early identification of risks and allow stakeholders to collaborate on finding effective solutions. It can also foster a culture of accountability, ensuring that all stakeholders take ownership of their responsibilities in risk management.

In conclusion, effective communication is crucial for successful risk management in construction projects. By implementing clear communication strategies, involving stakeholders, and maintaining transparency, project managers can mitigate risks and ensure the project's success.

Chapter 8: Integrating Risk Management into Project Planning

Risk management is a crucial aspect of construction project management, ensuring that potential risks are identified, assessed, and addressed in a timely and effective manner. However, implementing risk management practices can sometimes be seen as an additional step in the project planning process, adding unnecessary complexity and time. In this chapter, we will explore how risk management can be seamlessly integrated into project planning, and the impact it can have on the project schedule and budget. We will also discuss some best practices for effectively integrating risk management into project planning, ensuring a smooth and successful construction project.

Integrating Risk Management into Project Planning

Traditionally, risk management has been seen as a separate process from project planning, usually conducted after the project has been planned and budgeted. However, this approach can lead to delayed risk identification and mitigation, which can have a significant impact on the project schedule and budget. To avoid this, it is crucial to integrate risk management into the project planning process itself. The first step in integrating risk management into project planning is to ensure that all stakeholders involved in the project are aware of the importance of risk management. This includes project owners, contractors, subcontractors, and other team members. By creating a culture of risk awareness and acceptance, stakeholders can effectively identify and address risks throughout the project's lifecycle. Next, it is important to conduct a thorough risk assessment during the project planning stage. This involves identifying potential risks and evaluating their likelihood and impact on the project. By conducting this assessment early on in the planning process, risks can be addressed in a proactive manner, saving time and resources in the long run.

Impact of Risk Management on Project Schedule and Budget

Effective risk management can have a significant impact on the project's schedule and budget. By identifying and addressing risks early on in the project planning process, potential delays and cost overruns can be mitigated. This is especially important in construction projects, where delays can have a ripple effect on the entire project timeline.Moreover, by integrating risk management into project planning, risks can be factored into the project's budget, ensuring that adequate resources are allocated to address potential risks. This can also lead to cost savings, as risks are identified and addressed before they escalate into more significant and costly issues.

Risk Management Best Practices

To successfully integrate risk management into project planning, it is essential to follow some best practices. These include:

- Establishing a risk management team: A dedicated team with expertise in risk management should be responsible for identifying, assessing, and managing risks throughout the project.

- Regular risk reviews: Just as project progress is regularly reviewed, risks should also be periodically reassessed to ensure that potential risks have not been overlooked and that existing risks are being addressed effectively.

- Developing a risk management plan: This plan should outline the procedures and processes for managing risks, including risk identification, assessment, and response strategies.

- Utilizing risk management tools and software: Various software and tools are available that can help in identifying and managing risks more efficiently.

- Communication: Effective communication between all project stakeholders is crucial in identifying and addressing risks. It is essential to establish open communication channels and clearly communicate risk management responsibilities to each team member.

- Learning from past projects: Projects in the construction industry often face similar risks. By learning from past projects, organizations can identify common risks and develop strategies to effectively mitigate them in future projects.

In conclusion, integrating risk management into project planning is crucial for the successful execution of a construction project. By following best practices and involving all stakeholders in the risk management process, potential risks can be identified and addressed early on, leading to a smoother and more cost-effective project. From the initial planning stages to project completion, risk management should be a continuous process, ensuring the project's success and minimizing potential delays and cost overruns.

Chapter 9: Effective Risk Management Techniques for Construction Projects - Managing Project Budgeting and Cost Control

Managing project budgeting and cost control is an essential aspect of effective risk management in construction projects. The success of a project depends heavily on the ability to stay within the allocated budget and avoid cost overruns. In this chapter, we will discuss three key approaches to managing project budgeting and cost control: estimating and allocating contingency funds, managing cost overruns, and value engineering. Through these techniques, construction project managers can better manage risks and ensure the success of their projects.

Estimating and Allocating Contingency Funds

One of the primary ways to manage risk in project budgeting is by estimating and allocating contingency funds. Contingency funds are additional funds set aside in the project budget to cover unexpected costs or risks that may arise during construction. It is essential to accurately estimate the amount of contingency funds needed to mitigate potential risks without overestimating and affecting the project budget's overall feasibility. To estimate contingency funds, project managers should conduct a thorough risk assessment and identify potential risks that may impact the project budget. This can include factors such as changes in material costs, labor shortages, and delays in project timelines. By considering all possible risks and their potential impacts, project managers can determine an appropriate amount of contingency funds to allocate.It is crucial to allocate contingency funds effectively within the project budget. This means identifying which activities or aspects of the project may be more prone to risk and allocating a larger portion of contingency funds to those areas. Project managers should regularly review and update the contingency funds throughout the project to ensure they are properly allocated and available for any potential risks or unforeseen events.

Managing Cost Overruns

Despite careful planning and budgeting, cost overruns can still occur in construction projects. These are instances where the actual project costs exceed the allocated budget. Cost overruns can significantly impact a project's financial feasibility and potentially result in project failure if not managed effectively. To manage cost overruns, project managers must actively monitor the project's budget and expenses. Regularly reviewing and tracking project costs against the budget can help identify potential cost overruns early on, allowing for timely and effective decision-making. If cost overruns are identified, project managers should assess the cause and look for ways to mitigate and control the additional costs.Communication is key when managing cost overruns. Project managers must clearly communicate any cost overruns to the project team and stakeholders, including the reasons for the overruns and the proposed strategies for managing them. By being transparent and proactive in managing cost overruns, project managers can maintain trust and ultimately minimize the impact on the project's budget and timeline.

Value Engineering

Value engineering is a technique used to identify and reduce costs without compromising the project's overall quality and objectives. It involves analyzing every aspect of the project to determine if there are more cost-effective alternatives or ways to achieve the same results. Value engineering can be a powerful tool for managing project budgeting and cost control, as it allows project managers to identify potential cost savings and optimize the project's budget.Value engineering should be conducted at the early stages of a project to have the most significant impact on cost control. It requires a collaborative effort between project managers, engineers, and other members of the project team to review and analyze design and construction plans. Through brainstorming and evaluating alternatives, project teams can identify opportunities for cost savings and implement them into the project plan.

Conclusion

Effective risk management in construction projects requires proper management of project budgeting and cost control. Through estimating and allocating contingency funds, managing cost overruns, and value engineering, project managers can mitigate potential risks and ensure projects stay on budget. It is crucial to regularly review and adapt these techniques throughout the project's lifecycle to effectively manage any potential risks that may arise. By following these strategies, construction project managers can improve the chances of project success and ultimately achieve better outcomes.

Chapter 10: Contracts and Risk Management

Importance of Clear Contracting

When it comes to managing risk in construction projects, clear and well-defined contracts are essential. Contracts serve as a legally binding agreement between all parties involved in a project and outline the rights, responsibilities, and obligations of each party. A poorly written or vague contract can lead to misunderstandings, disputes, and ultimately, increased risk for all parties involved. In order to minimize risk, it is crucial for construction companies to ensure that their contracts are clear, concise, and cover all necessary aspects of the project. This includes outlining the scope of work, project timeline, payment terms, change order procedures, and any special conditions or requirements. By clearly defining these aspects in a contract, there is less room for misinterpretation or disputes to arise.Furthermore, clear contracting also establishes a shared understanding and expectations among all parties involved. This can help to prevent delays, cost overruns, and other risks that may occur due to miscommunication. By having a well-defined contract in place, all parties can refer back to it throughout the project and ensure that everyone is on the same page.

Legal Considerations for Risk Management

In addition to ensuring clear contracts, it is also important for construction companies to consider the legal aspects of risk management. Construction projects are subject to a wide range of local, state, and federal laws and regulations, and failure to comply with these laws can result in legal consequences and increased risk. One key legal consideration for risk management is insurance. Construction companies should have adequate insurance coverage to protect against potential risks and liabilities. This may include general liability insurance, professional liability insurance, and workers' compensation insurance. Adequate insurance coverage not only helps to minimize risk, but it can also protect the company from financial losses in the event of a claim or lawsuit.Another important legal consideration for risk management is ensuring that all contracts and agreements are compliant with applicable laws and regulations. This includes ensuring that all necessary permits and licenses are obtained, and that the project adheres to all building codes, zoning laws, and environmental regulations.

Failure to comply with these legal requirements can not only result in penalties and fines, but it can also pose serious risks to the project and the safety of workers and the surrounding community.

Mitigating Risks Through Contracts

One of the primary purposes of contracts is to allocate and mitigate risks among project parties. By clearly outlining responsibilities and obligations, contracts can help to identify potential risks and determine who is responsible for managing them. One common way to mitigate risks through contracts is through the use of indemnification clauses. These clauses indemnify one party from any loss, damage, or liability that may result from the actions or negligence of the other party. For example, construction companies may include an indemnification clause in their contracts with subcontractors, placing the responsibility on the subcontractor to cover any costs or damages resulting from their work. Another way to mitigate risks through contracts is through the inclusion of specific warranties and guarantees. These provide assurance to clients that the construction company is responsible for correcting any defects or issues with the project. By providing these warranties in contracts, the construction company is taking on the risk of potential defects, rather than the client. Effective risk management through contracts also involves carefully reviewing and negotiating the terms of contracts. This includes making sure that contracts are properly reviewed by legal counsel to ensure they are compliant with applicable laws and regulations. It is also important to carefully consider the language used in contracts and ensure that it is clear and unambiguous.In addition, construction companies should also have a solid understanding of the risks involved in a project and negotiate contract terms accordingly. For high-risk projects, companies may want to include more protective clauses in contracts, while for lower-risk projects, they may be able to take on more liability and risk.

Conclusion

Clear and well-defined contracts are essential for effective risk management in construction projects. By ensuring contracts are clear, legally compliant, and properly negotiated, construction companies can minimize misunderstandings and disputes, allocate and mitigate risks, and protect themselves from potential liabilities and losses. It is worth investing time and resources into drafting and reviewing contracts to ensure they effectively manage risks and contribute to the overall success of a project.

Chapter 11: Safety and Risk Management

Safety and risk management go hand in hand in the construction industry. A project cannot be deemed successful without ensuring the safety of the workers, the public, and the environment. Safety risks can have a major impact on a project, causing delays, accidents, and even fatalities. It is the responsibility of every construction company to have a strong safety culture and implement effective risk management techniques to mitigate safety risks. In this chapter, we will explore the relationship between safety and risk management, as well as ways to identify and mitigate safety risks, and ensure compliance with safety regulations.

Relationship Between Safety and Risk Management

The main goal of risk management in construction is to identify and mitigate potential risks before they can affect the project. This includes safety risks. An effective risk management plan should not only focus on project completion within budget and schedule, but also on ensuring the safety of everyone involved in the project. By addressing safety risks proactively, a company can avoid costly delays, injuries, and fatalities, which can ultimately lead to a negative reputation and legal consequences.Furthermore, safety risks can also lead to financial consequences, such as fines and compensation costs. A company may also experience difficulties in obtaining insurance or funding if they have a poor safety record. This highlights the close relationship between safety and risk management and why it is essential to address safety risks in a comprehensive risk management plan.

Identifying and Mitigating Safety Risks

The first step in managing safety risks is to identify them. This involves conducting a thorough risk assessment and analysis to determine potential hazards and their likelihood of occurring. A risk assessment can be done at the beginning of a project, but it should also be an ongoing process, with risks regularly reviewed and updated throughout the project.Once safety risks have been identified, the next step is to mitigate them. This includes implementing measures to eliminate the risk or reduce the likelihood of it occurring. This may involve implementing safety protocols, providing proper training to workers, providing personal protective equipment, and conducting regular safety inspections and audits. It is crucial to involve all stakeholders, including workers, in the risk mitigation process to ensure their buy-in and to effectively address potential safety risks.

Ensuring Compliance with Safety Regulations

In addition to identifying and mitigating safety risks, it is also crucial to ensure compliance with safety regulations. Construction companies must stay up to date with all safety regulations and standards, as well as any changes or updates that may occur. This may include OSHA regulations, local building codes, and industry-specific safety standards. Proper documentation and record-keeping are also essential to ensure compliance. This includes documenting safety training, safety meetings, and safety inspections. By keeping accurate records, companies can demonstrate their commitment to safety and compliance in the event of any legal issues. In addition to following safety regulations, it is also important to regularly review and update safety protocols to ensure they are in line with current best practices and regulations. This may involve conducting regular safety audits and involving workers in the process to get their feedback and suggestions for improvement.

In conclusion, safety and risk management are two sides of the same coin in the construction industry. Without proper safety measures and risk management techniques, a project is at risk of delays, accidents, and financial consequences. By understanding the relationship between safety and risk management, identifying and mitigating safety risks, and ensuring compliance with safety regulations, construction companies can create a safer working environment and ensure the success of their projects.

Chapter 12: Human Resource Management and Risk

Impact of Human Factors on Risk

Human resources are a crucial component of any construction project. It is the people who design, plan, and execute the project, making their contribution vital to its success. However, human factors can also pose a significant risk to the project if not managed effectively. The impact of human factors on risk in construction projects should not be underestimated, as it can lead to delays, cost overruns, and even accidents. One of the most significant human factors that can affect risk in construction projects is communication. Poor communication within the project team can lead to misunderstandings, delays, and mistakes. Ineffective communication with stakeholders and subcontractors can also result in delays and misunderstandings, potentially leading to disputes. It is essential to establish clear and open lines of communication within the project team and with external stakeholders to reduce communication-related risks. Another crucial human factor that can impact risk is the competency of project team members. Lack of knowledge and skills can lead to errors in design, planning, and execution, resulting in rework and delays. It is crucial to ensure that all project team members have the necessary training and qualifications for their roles to minimize the risk of human error. The working culture and attitudes of the project team members also play a significant role in risk management. Team members who are resistant to change, have a negative attitude, or are not committed to the project can create a toxic work environment and increase the likelihood of errors and delays. It is crucial to promote a positive and collaborative working culture within the project team to mitigate the risks associated with negative attitudes and lack of commitment.

Training and Development for Risk Management

As stated earlier, competency is a crucial factor in managing risks in construction projects. This makes training and development a vital component of risk management. Investing in training and developing the skills of the project team members can significantly reduce the risk of human error, delays, and accidents. One of the ways to ensure that project team members are adequately trained is by conducting a thorough analysis of their current skills and identifying any gaps. This can help determine the specific training and development needs of each team member. It is also essential to provide ongoing training and development opportunities to keep the project team up-to-date with new technologies, processes, and best practices. In addition to technical skills, it is also crucial to provide training in risk management strategies and techniques. This can help project team members identify potential risks and take proactive measures to mitigate them. It is also essential to train team members on effective communication and teamwork skills to promote a positive working culture and reduce the risks associated with human factors.

Managing Human Error

Despite the best training and development efforts, human error can still occur in construction projects. An effective risk management strategy should include measures to mitigate the impact of human error on the project. This can include implementing quality control procedures, conducting thorough reviews and inspections of design and plans, and implementing a system for reporting and addressing errors promptly. It is also crucial to create a culture of accountability and continuous improvement within the project team. Team members should feel comfortable reporting errors without fear of retribution. This can help identify potential risks and prevent them from causing significant issues in the project. In addition to managing human error within the project team, it is also essential to consider subcontractors and suppliers. It is crucial to have clear communication and accountability guidelines in place to ensure that all project partners are aware of their responsibilities and the potential consequences of human error.

In conclusion, human resources play a crucial role in risk management in construction projects. The impact of human factors on risk cannot be ignored, and it is essential to effectively manage these factors to ensure project success. This includes investing in training and development, fostering a positive working culture, and having strategies in place to manage and mitigate human error.

Chapter 13: Quality Management and Risk

Quality Control Measures to Minimize Risks

Quality management is an integral part of any construction project, and it plays a crucial role in minimizing risks. By implementing effective quality control measures, construction companies can prevent costly mistakes, delays, and potential hazards. Quality control is not just about meeting standards, but it is also about ensuring that the final product meets the client's expectations and specifications. One of the key quality control measures to minimize risks is setting clear quality objectives. This involves defining project requirements, specifications, and standards. Without clear objectives, quality control becomes challenging, and risks may arise due to confusion and lack of understanding. Another important measure is quality assurance, which involves the systematic and planned auditing of materials, processes, and equipment to ensure they meet the required standards. This helps in identifying potential risks early on, allowing for corrective action to be taken before they become major issues.Document control is also crucial in quality management as it ensures that all project-related documents, such as contracts, plans, and specifications, are up to date and easily accessible. This reduces the risk of using outdated or incorrect information, which can lead to errors and delays.

Risk of Defects and Delays

One of the biggest risks in construction projects is the occurrence of defects and delays. Defects, whether in the materials, design, or workmanship, can be costly to rectify and may even compromise the safety of the structure. Delays, on the other hand, can result in additional expenses, loss of productivity, and damage to the reputation of the company. To minimize the risk of defects, quality control measures such as proper material testing, inspection of work, and adherence to specifications and standards must be implemented. These measures help in identifying and correcting defects before they escalate into major issues.On the other hand, to mitigate the risk of delays, construction companies must have proper project planning and scheduling in place. This involves setting realistic timelines, considering potential delays, and having contingency plans in place. Effective communication and collaboration between all parties involved in the project are also crucial in preventing delays.

Ensuring Quality in Construction Projects

Ensuring quality in construction projects goes beyond just meeting standards and specifications. It requires a proactive approach that involves identifying and addressing potential risks before they become major issues. One way to ensure quality is to involve all stakeholders in the quality management process. This includes the client, designers, contractors, and subcontractors. By involving all parties, there is a shared understanding of project requirements, expectations, and potential risks, leading to a higher likelihood of project success. Regular inspections and quality audits throughout the project are also essential in ensuring quality. This involves conducting physical inspections of the work, reviewing documentation, and verifying compliance with standards and specifications. Any issues identified must be promptly addressed to prevent them from escalating into major risks. Moreover, continuous training and education of all project team members on quality management techniques and best practices can significantly contribute to the success of a construction project. This ensures that all team members are equipped with the necessary knowledge and skills to identify and address potential risks to quality.

In conclusion, quality management is an essential aspect of managing risks in construction projects. By implementing effective quality control measures, minimizing the risk of defects and delays, and ensuring quality throughout the project, construction companies can significantly reduce the likelihood of costly mistakes, delays, and potential hazards.

Chapter 14: Identifying and Managing Environmental Risks

The construction industry has a significant impact on the environment, and as such, it is crucial for construction projects to identify and manage potential environmental risks. Environmental risks can include pollution, land and water contamination, and harm to wildlife and ecosystems. In this chapter, we will explore the importance of identifying and managing environmental risks in construction projects.

Importance of Adhering to Environmental Regulations

One of the main reasons why it is essential to identify and manage environmental risks in construction projects is to adhere to environmental regulations. Governments around the world have implemented various laws and regulations to protect the environment and ensure that construction projects do not have a detrimental impact on the ecosystem.By not adhering to these regulations, construction projects not only put the environment at risk but also face legal consequences. These consequences can include fines, project delays, and even cancellation of the project. Therefore, it is imperative for construction projects to identify and manage environmental risks to avoid any legal complications and ensure compliance with regulations.

Environmental Risk Management Process

The first step in managing environmental risks in construction projects is to identify potential risks. This involves conducting a thorough site assessment to determine any potential hazards and their potential impact on the environment. A comprehensive site assessment may include soil and water testing, analysis of the surrounding flora and fauna, and identifying any historical contamination.Once potential risks have been identified, the next step is to assess the level of risk and determine the appropriate mitigation measures. This process should involve input from environmental experts to ensure that the best possible course of action is taken.

Use of Sustainable Practices

Another crucial aspect of managing environmental risks in construction projects is incorporating sustainable practices. These practices not only reduce the negative impact on the environment but also have long-term cost-saving benefits. Sustainable practices can include using eco-friendly materials, reducing waste and emissions, and implementing energy-efficient measures.Many government organizations and private companies have also started incorporating sustainability requirements into their procurement processes. This means that construction companies must adhere to sustainable practices to win bids and contracts, making it necessary for construction projects to identify and manage environmental risks.

Benefits of Proper Environmental Risk Management

Effective environmental risk management can have many benefits for construction projects. These benefits include:

- Avoiding legal complications and costly fines

- Enhancing the company's reputation as a responsible and environmentally-conscious organization

- Increased chances of winning bids and contracts from organizations with sustainability requirements

- Reduced long-term costs due to implementing sustainable practices

- Preserving the environment for future generations

Involving Stakeholders in Environmental Risk Management

Stakeholder involvement is vital in managing environmental risks in construction projects. It is essential to involve stakeholders such as local communities, environmental experts, and government organizations in the decision-making process. This not only helps in identifying potential risks but also ensures that all concerns and perspectives are taken into account.Involving stakeholders also helps educate and raise awareness about the importance of protecting the environment and encourages collective responsibility in managing environmental risks.

Conclusion

In conclusion, identifying and managing environmental risks in construction projects is crucial for a variety of reasons. It allows for compliance with environmental regulations, promotes sustainability, and has several long-term benefits. By involving stakeholders and incorporating sustainable practices, construction projects can reduce their impact on the environment and contribute to a more environmentally-friendly industry. Companies that prioritize environmental risk management can also enhance their reputation and improve their chances of success in the market.

Chapter 15: Understanding and Managing supply chain risks

Understanding Supply Chain Risks

A construction project involves a complex network of suppliers, subcontractors, and other stakeholders. This network is known as the supply chain. It plays a crucial role in the success of a project, but it also brings along various risks that need to be identified and managed effectively. Understanding supply chain risks is essential for any construction project to be completed on time, within budget, and with high quality.Supply chain risks can include delays in material deliveries, quality issues, financial risks, and even environmental and social risks. These risks can ultimately affect the project's progress, leading to delays and cost overruns. Therefore, it is crucial to have a thorough understanding of these risks and implement strategies to manage them effectively.To understand supply chain risks, it is essential to have a clear understanding of the entire supply chain process. This includes identifying all the stakeholders involved, mapping out the flow of materials and information, and assessing their potential impact on the project. This process can be complex, but it is crucial for effective risk management.

Managing Risks in the Supply Chain

Once the supply chain risks have been identified, it is vital to have a plan in place to manage them. This involves a proactive and continuous effort to mitigate and monitor risks throughout the project's lifespan. Here are some strategies for managing risks in the supply chain:

1. Develop a risk management plan: A detailed risk management plan should be developed for the entire supply chain process. This plan should include all identified risks, their potential impact, and the strategies for mitigating and monitoring them.

2. Partner with reliable suppliers: A strong and reliable supplier network is the

foundation of an effective supply chain. Partnering with suppliers that have a good track record and are committed to quality and timely delivery can reduce supply chain risks significantly.

3. Conduct regular audits: Regular audits of suppliers and subcontractors' performance can help identify potential risks and implement corrective actions before they turn into a problem.

4. Communicate effectively: Effective communication among all stakeholders is crucial for managing risks in the supply chain. Regular and transparent communication can help identify and address potential issues before they escalate.

5. Have contingency plans in place: Despite the best efforts, some risks may still occur. Having contingency plans in place can help minimize the impact of these risks on the project's progress.

Building Resilient Supply Chain Partnerships

In addition to managing risks, it is also essential to build resilient partnerships within the supply chain. These partnerships go beyond the traditional supplier-contractor relationship and involve trust, open communication, and collaboration. Building such partnerships can bring long-term benefits, including reduced risks and increased operational efficiency.Here are some tips for building resilient supply chain partnerships:

1. Develop a shared vision: A shared vision is critical for building strong partnerships within the supply chain. All stakeholders should have a clear understanding of the project's goals and work towards achieving them together.

2. Invest in relationships: Developing strong relationships with suppliers, subcontractors, and other stakeholders takes time, effort, and investment. Investing in relationships can bring long-term benefits, including increased trust, open communication, and collaboration.

3. Collaborate on risk management: Effective risk management requires collaboration and cooperation from all stakeholders. Involve them in identifying and managing risks, and develop contingency plans together.

4. Monitor and evaluate performance: Regularly monitoring and evaluating the performance of suppliers and subcontractors can help identify any potential issues and address them promptly.

5. Celebrate successes: Celebrate milestones and successes with all stakeholders involved in the supply chain. This can help build a sense of teamwork and appreciation, leading to stronger partnerships.

Building resilient supply chain partnerships is not an easy task, but the benefits are worth the effort. These partnerships can reduce risks, improve project outcomes, and even lead to cost savings.

In conclusion, understanding supply chain risks, and effectively managing them is crucial for the success of any construction project. It requires a proactive and collaborative effort from all stakeholders, as well as continuous monitoring and evaluation. Through the strategies mentioned, construction projects can build resilient supply chain partnerships and ensure the smooth execution of the project.

Chapter 16: Technology and Risk Management

Role of Technology in Risk Management

Technology has completely revolutionized our daily lives, and the construction industry is no exception. With the advancement of technology, we now have access to tools and resources that were once unimaginable. One of the key areas where technology is making a significant impact is in risk management. The role of technology in risk management for construction projects is to assist project managers in identifying, assessing, and mitigating potential risks. By streamlining processes and providing timely and accurate data, technology is changing the way construction companies approach risk management.

Tools and Software for Risk Management

Gone are the days of maintaining spreadsheets and manual paperwork for risk management. Technology has brought a plethora of tools and software that help construction companies manage risks efficiently and effectively. These tools and software offer various features and capabilities, including data tracking, risk identification, risk analysis, and risk response planning. With real-time updates and reminders, these tools allow project managers to have a better understanding of potential risks and take proactive measures to mitigate them. One such tool is Building Information Modeling (BIM), which has become increasingly popular in the construction industry. BIM integrates data and information from multiple sources, giving project managers a comprehensive view of the project and potential risks involved. It allows for better visualization, collaboration, and communication among project team members, facilitating risk management. With the ability to simulate and test different scenarios, project managers can identify potential risks and take necessary precautions before the construction process even begins. Another valuable tool for risk management is risk management software. These software solutions offer customized risk management plans, project tracking, and reporting capabilities. With the ability to automate repetitive tasks and provide real-time updates, these software solutions save time and effort for project managers, allowing them to focus on other critical aspects of the project.

Integrating Technology into Risk Management Plans

Integrating technology into risk management plans is crucial for ensuring successful project outcomes. By leveraging technology tools and software, risk management can be incorporated into every stage of the project, allowing for better risk identification, assessment, and response planning. With the availability of real-time data and analytics, project managers can make informed decisions and implement necessary changes to reduce potential risks. One of the key benefits of integrating technology into risk management plans is improved communication and collaboration. With the use of technology, project team members can easily share information, documents, and updates, eliminating communication barriers and ensuring that everyone is on the same page. By promoting transparency and collaboration, technology helps in minimizing misunderstandings and conflicts, which can be potential risks for the project. Furthermore, technology also allows project managers to track progress and identify deviations from the initial risk management plan. With the use of data analytics, project managers can identify trends and patterns in risks, enabling them to make adjustments to their plan accordingly. This proactive approach to risk management helps in minimizing the impact of risks and ensuring project success.

In conclusion, technology plays a critical role in risk management for construction projects. With its advanced tools and software, it streamlines processes, improves communication and collaboration, and provides timely and accurate data. By incorporating technology into risk management plans, project managers can identify potential risks and take necessary measures to mitigate them, ensuring successful project outcomes.

Chapter 17: Case Studies of Successful Risk Management in Construction Projects

Risk management is an essential aspect of any construction project. It involves identifying potential risks and developing strategies to mitigate their impact on the project's timeline, budget, and quality. While no construction project is entirely free of risks, effective risk management techniques can significantly reduce the likelihood and severity of potential issues.In this chapter, we will explore some real-world examples of successful risk management in construction projects. These case studies will highlight the importance of implementing proper risk management techniques and the resulting benefits for the project.

Case Study 1: The Burj Khalifa Project

The Burj Khalifa is the tallest building in the world, standing at an impressive 2,722 feet tall. Its construction was a massive undertaking that faced numerous challenges, including adverse weather conditions, a tight timeline, and a complex design. Despite these potential risks, the project was completed successfully, and the building is now an iconic landmark in Dubai. One of the key factors contributing to the project's success was the implementation of effective risk management techniques. The project team conducted extensive risk assessment and developed strategies to mitigate potential risks. This included contingency planning for unfavorable weather conditions, implementing strict quality control measures, and regular communication and decision-making processes.As a result, the Burj Khalifa project was completed on time and within budget, with minimal disruptions and safety incidents. It serves as an excellent example of how proactive risk management can ensure the success of even the most challenging construction projects.

Case Study 2: The Olympic Stadium in London

The construction of the Olympic Stadium in London for the 2012 Summer Olympics faced various risks, including tight deadlines, a limited budget, and strict sustainability requirements. To ensure the project's success, the project team adopted an integrated risk management approach, involving all stakeholders throughout the project's lifecycle. This approach involved identifying and analyzing potential risks, developing contingency plans, and regularly reviewing and updating risk management strategies. It also involved engaging with the local community and implementing rigorous safety protocols to mitigate any potential risks to the public. As a result, the Olympic Stadium was completed on time and within budget, meeting all sustainability goals and without any major incidents. This successful outcome would not have been possible without a robust risk management plan in place.

Common Mistakes and How to Avoid Them

Mistake #1: Not Conducting Proper Risk Assessments

One of the most common mistakes in risk management is not conducting thorough risk assessments. Without a comprehensive understanding of potential risks, it is challenging to develop effective risk management strategies. This can lead to costly delays, budget overruns, and quality issues during the project. To avoid this mistake, it is crucial to involve all stakeholders in the risk assessment process, including designers, contractors, and even clients. This will allow for a comprehensive identification and analysis of potential risks from all perspectives.

Mistake #2: Failure to Update Risk Management Strategies

Risk management is not a one-time process; it requires continuous monitoring and updating throughout the project's lifecycle. Failing to update risk management strategies as the project progresses can lead to unforeseen issues and disruptions down the line. To avoid this mistake, it is essential to regularly review and update risk management strategies, taking into account any changes in project scope, timeline, or budget. This will ensure that the project team is adequately prepared to deal with any potential risks that may arise.

Mistake #3: Lack of Communication

Effective communication is key to any successful construction project, and it is especially crucial in risk management. Failure to communicate potential risks and their mitigation strategies can result in misunderstandings and delays, ultimately affecting the project's success.To avoid this mistake, it is essential to establish clear communication channels and protocols, ensuring that all stakeholders are kept informed of any potential risks and the project's progress. Regular communication will also allow for prompt action in the event of any unforeseen issues, minimizing their impact on the project.

Mistake #4: Ignoring Lessons Learned

Every construction project is a learning experience, and it is vital to incorporate these lessons into future risk management strategies. Ignoring lessons learned can result in repeating the same mistakes, leading to avoidable delays and disruptions.To avoid this mistake, it is crucial to conduct thorough post-project reviews and document any lessons learned. It is also essential to share these findings with the project team and incorporate them into future risk management plans.

Mistake #5: Not Involving Experts

Construction projects are complex, and they often require specialized expertise to identify and mitigate potential risks effectively. Not involving subject matter experts in the risk management process can lead to overlook of crucial risks and inadequate strategies in place. To avoid this mistake, it is essential to involve experts, such as risk consultants or legal advisors, in the risk management process. This will ensure a comprehensive assessment of potential risks and the development of appropriate mitigation strategies.

In conclusion, effective risk management techniques are critical to the success of any construction project. By learning from successful case studies and avoiding common mistakes, project teams can ensure the smooth and timely completion of their projects. Remember, proactive risk management is always the key to success.

Chapter 18: International Perspectives on Risk Management

Risk management is a crucial aspect of any construction project, and it becomes even more critical when working on an international scale. As we continue to see increased globalization, construction companies are taking on projects in various parts of the world, facing cultural differences, legal regulations, and unique risks. Managing these risks effectively can make the difference between a successful project and a failure. In this chapter, we will dive into the cultural differences in risk management and the best practices for international risk management.

Cultural Differences in Risk Management

One of the biggest challenges when it comes to international risk management is understanding and navigating through cultural differences. When working on a project in a different country, a construction company must adapt to the local culture, which can have a significant impact on risk management. Every country has its own beliefs, values, and ways of doing business, which can greatly affect how risks are perceived and managed. For instance, in Western cultures, risk management is often seen as an analytical process, where risks are carefully identified and assessed, and strategies are developed to mitigate them. However, in some Eastern or African cultures, risk management may not have the same level of importance, and risks may not be adequately addressed until they become an issue. This can lead to misunderstandings and communication breakdowns between project teams, resulting in delays and increased costs. Another cultural difference that can affect risk management is the level of hierarchy in a society. In some countries, decision-making and risk management are centralized, and approval from higher authorities is required for any changes or deviations from the original plan. This can cause delays and disruptions in the project if not accounted for in the risk management plan.Language barriers, different time zones, and varying work ethics can also present challenges in effectively managing risks in an international project. It is crucial for construction companies to understand and address these cultural differences to ensure effective risk management and successful project delivery.

Best Practices in International Risk Management

With the increased globalization of the construction industry, there has been a growing recognition of the need for best practices in international risk management. These practices help construction companies identify, assess, and manage risks more effectively, regardless of cultural differences or geographical barriers.

1. Establish Communication Protocols: Communication is key in a successful international project, especially when it comes to risk management. Construction companies must establish clear communication protocols with all project team members, including stakeholders, contractors, and local authorities. This ensures timely communication and reduces the chances of misunderstandings.

2. Conduct Thorough Risk Assessments: A thorough risk assessment is essential for identifying and understanding the unique risks associated with a project in a different country. This includes cultural, economic, political, and environmental factors that may affect the project. Construction companies should also consider engaging with local experts to gain a better understanding of the risks involved in the project.

3. Develop a Comprehensive Risk Management Plan: A well-developed risk management plan should be tailored to the specific project and its risks. This plan should include strategies for risk mitigation, contingency plans, and a communication plan. It should also outline the roles and responsibilities of all project team members regarding risk management.

4. Study Local Laws and Regulations: When working on an international project, construction companies must be well-versed in the local laws and regulations that may affect the project. This includes building codes, safety standards, and legal requirements for construction permits. Non-compliance with local laws and regulations can result in costly delays and potential legal issues.

5. Build Strong Relationships: Developing strong relationships with local stakeholders and project team members is crucial for successful risk management. Understanding and respecting cultural differences can help foster trust and cooperation, leading to more effective risk management.

6. Monitor and Review Risks Regularly: Risk management is an ongoing process, and risks can change throughout the duration of a construction project. Therefore, it is

essential to regularly review and monitor risks to ensure that the risk management plan is still relevant and effective.

In conclusion, managing risks in an international construction project requires a deep understanding of cultural differences and the implementation of best practices. By establishing clear communication, conducting thorough risk assessments, and developing a comprehensive risk management plan, construction companies can mitigate risks and ensure the successful delivery of their projects in any part of the world.

Chapter 19: Legal and Ethical Considerations

Effective risk management is vital for the success of any construction project. However, it is not just a matter of following best practices and implementing the right strategies. As with any business practice, there are legal and ethical considerations that must be taken into account when managing project risks.

Legal Obligations for Risk Management

Before embarking on a construction project, it is important to understand the legal obligations regarding risk management. This includes complying with local, state, and federal laws, as well as any regulations set forth by industry governing bodies. One of the main legal obligations for risk management in construction projects is to ensure the safety of workers and the general public. This is typically achieved through compliance with Occupational Safety and Health Administration (OSHA) regulations. These regulations cover everything from ensuring proper training and protective gear for workers to maintaining safe working conditions on the construction site. In addition to safety requirements, there may also be legal requirements for risk management related to environmental protection. Depending on the project and location, there may be regulations for waste disposal, air and water quality, and protection of natural resources. It is important to consult with environmental experts and ensure compliance with these regulations to avoid potential legal issues. Another legal consideration for risk management is potential liability for damages or injuries. This is why it is crucial to have proper insurance coverage and contracts in place. Contracts should clearly outline the responsibilities and liabilities of all parties involved in the project, and insurance coverage should be comprehensive enough to address any potential risks that may arise.

Ethical Principles in Managing Project Risks

In addition to legal obligations, there are also ethical principles that should guide risk management in construction projects. First and foremost, a commitment to honesty and transparency is essential in ensuring ethical practices. This includes being upfront about potential risks and having open communication with all stakeholders. Integrity is also a key ethical principle in risk management. This means acting with honesty and adhering to ethical standards and professional codes of conduct. It also involves taking responsibility for mistakes and making ethical decisions even when they may not be in the best financial interest for the company.Respect for individuals and their rights is another important ethical consideration. This includes respecting the rights of workers and respecting the rights of the community in which the construction project is taking place. Risk management strategies should take into account the potential impacts on the local community and work to mitigate any negative effects.

Dealing with Conflict of Interest

Construction projects often involve multiple parties with different interests and motivations. This can create potential conflicts of interest, which can complicate risk management efforts. It is crucial to identify and address any conflicts of interest at the beginning of a project to avoid potential issues down the line. Transparency and clear communication are key in dealing with conflicts of interest. All parties involved should be aware of any potential conflicts and efforts should be made to mitigate them. This may involve implementing independent oversight or creating clear guidelines for decision making. In some cases, it may be necessary to involve legal counsel to address conflicts of interest. It is important to have a robust conflict resolution process in place to handle any disputes that may arise. This process should be fair and transparent to all parties involved.By understanding and adhering to legal obligations and ethical principles, as well as effectively managing conflicts of interest, construction companies can ensure the success of their risk management efforts and maintain a positive reputation in the industry.

Closing Thoughts

In conclusion, risk management in construction projects is not just about implementing effective strategies and techniques – it also involves taking into account legal and ethical considerations. By understanding and complying with legal obligations, adhering to ethical principles, and effectively managing conflicts of interest, construction companies can mitigate potential risks and ensure the success of their projects.

Chapter 20: Managing Crisis Risks in Construction Projects

Understanding and Preparing for Crisis Situations

In the construction industry, unforeseen events and emergencies can occur at any time, which can greatly impact the progress and success of a project. These crisis situations can range from natural disasters, such as hurricanes or earthquakes, to accidents on the job site or financial problems. As a project manager, it is crucial to understand and prepare for these potential crisis risks in order to mitigate their impact on the project. The first step in understanding crisis situations is to conduct a thorough risk assessment. This involves identifying potential risks and their likelihood of occurring, as well as evaluating the potential consequences if they were to happen. By doing so, you can prioritize and focus on the most critical risks that can have the greatest impact on the project.Next, it is important to have a crisis management plan in place. This plan should outline procedures and protocols for handling crisis situations, as well as identify roles and responsibilities of team members in the event of a crisis. It should also include a communication plan for keeping all stakeholders informed and updated throughout the crisis.

Crisis Communication

Effective communication is crucial in managing crisis situations in construction projects. It is important to establish clear lines of communication and designate a spokesperson for the project. This person should be trained in crisis communication and have the ability to remain calm and handle difficult situations under pressure. In addition, it is important to have a communication plan in place for both internal and external stakeholders. This can include regular updates through email, phone calls, or meetings, as well as a designated point of contact for stakeholders to reach out to with any concerns or questions.Transparency is key in crisis communication. It is important to be honest and upfront about the situation and its potential impact on the project. By being transparent, trust can be built with stakeholders and credibility can be maintained.

Strategies for Managing Crisis Risks

Prevention is always better than cure when it comes to managing crisis risks. By identifying potential risks and putting measures in place to prevent them, the impact of a crisis can be greatly reduced. One way to prevent crisis risks is to have a well-trained and competent project team. This means investing in proper training for team members and ensuring they have the necessary skills and knowledge to handle unexpected situations. In addition, having contingency plans in place can also help mitigate the impact of a crisis. This involves creating alternative strategies and backup plans in case a crisis does occur. These plans should be regularly reviewed and updated to ensure their effectiveness. Collaboration with stakeholders is another important strategy for managing crisis risks. By involving all stakeholders in the risk assessment and crisis management planning process, potential risks can be identified and managed more effectively. This also helps to build trust and credibility with stakeholders, as they feel involved and informed about the project.Finally, it is important to remain adaptable and flexible in the face of a crisis. Plans may need to be adjusted and new strategies may need to be implemented in order to effectively manage the situation. Having an open-minded and solution-focused mindset can help in navigating through a crisis and minimizing its impact on the project.

In conclusion, managing crisis risks is a crucial aspect of effective risk management in construction projects. By understanding and preparing for potential crisis situations, having a clear crisis communication plan, and implementing strategies for managing crisis risks, project managers can effectively mitigate the impact of these events. As the saying goes, "Hope for the best, but prepare for the worst." By taking proactive measures, construction projects can continue to progress and succeed even in the face of unexpected challenges and crisis situations.

Chapter 21: The Importance of Design in Risk Management

Design plays a critical role in the success of any construction project. It involves not only the aesthetic aspect but also the functional and safety considerations. In the context of risk management, design is often seen as the first line of defense against potential risks. Each design decision has the potential to either increase or decrease the likelihood and impact of risks on the project. In this chapter, we will explore the role of design in risk management, how to identify and mitigate design risks, and the importance of designing for resilience.

Role of Design in Risk Management

Design is not only about creating visually pleasing structures, but it also involves creating a safe, efficient, and functional environment for the end-users. Therefore, design has a significant impact on the overall risk profile of a construction project. A well-thought-out design can greatly reduce potential risks and increase the likelihood of project success. One of the primary ways in which design influences risk management is through the identification and mitigation of potential risks. Design professionals are trained to consider all aspects of a project and anticipate potential challenges that may arise during construction. By incorporating risk management into the design process, designers can proactively address potential risks before they become major issues.Moreover, design also plays a role in cost control and budget management. Poor design decisions can lead to costly changes and delays, which can have a significant impact on a project's budget and schedule. By considering risk management in the design phase, designers can help minimize the potential for costly changes during construction.

Identifying and Mitigating Design Risks

During the design phase, it is crucial to identify potential risks and develop mitigation strategies. This involves not only considering the physical aspects of design but also the human and environmental factors that may impact the project. One effective way to identify design risks is through a thorough analysis of the project requirements. This includes understanding the client's needs, the site conditions, budget constraints, and any regulatory or environmental requirements. By thoroughly understanding these factors, designers can identify potential risks and develop design solutions that mitigate them. Another important aspect of mitigating design risks is through effective communication and collaboration. Design professionals should work closely with project stakeholders, including the client, contractors, and engineers, to ensure that all risks are considered and adequately addressed. Collaboration can also help identify potential conflicts and challenges early on, allowing for effective risk management and resolution.

Designing for Resilience

In recent years, there has been a growing emphasis on designing for resilience in the construction industry. Resilient design focuses on creating structures and systems that can withstand and bounce back from unexpected events or challenges. By incorporating resilience into design, projects can better mitigate and manage risks, ultimately leading to greater project success. One key aspect of designing for resilience is incorporating sustainability into the design process. Sustainable design focuses on creating structures that have a minimal impact on the environment and can adapt to changing conditions. This approach not only reduces environmental risks but also helps future-proof projects against potential challenges such as climate change. In addition to sustainability, designing for resilience also involves considering the end-users and their needs. Design professionals should strive to create structures that can withstand potential physical or social disruptions and provide a safe and functional environment for the end-users.

Conclusion

The design phase is crucial in risk management for construction projects. It sets the foundation for success by identifying and mitigating potential risks, ensuring cost and budget control, and incorporating resilience into the project. By considering the role of design in risk management, construction projects can be better equipped to handle potential challenges and increase the likelihood of project success. Collaboration, effective communication, and a focus on sustainability and resilience are key to successfully managing risks through the design process.

Chapter 22: Effective Risk Management in Different Project Delivery Methods

Comparison of Different Delivery Methods

In the world of construction, there are various project delivery methods that have been developed over the years. Each method has its own unique characteristics, advantages, and drawbacks. When it comes to risk management, it is crucial to understand the differences between these methods and how they can affect the overall success of a construction project. The traditional method of project delivery, known as the design-bid-build method, involves the sequential phases of design, bidding, and construction. This method provides a clear separation of roles and responsibilities, however, it also leads to a lack of collaboration and communication between parties, which can increase the risk of errors and delays. Another popular method is the design-build method, where the client contracts with a single entity for both design and construction. This method has the potential to reduce project duration and increase collaboration, but it can also result in a lack of transparency and accountability. In contrast, the construction manager at risk (CMAR) method involves the contractor being selected during the design phase, allowing them to provide input on constructability, cost, and schedule. While this method promotes collaboration and risk assessment, it also requires a high level of trust between the client and contractor.Lastly, the design-build-operate (DBO) method involves the contractor being responsible for the design, construction, and operation of the project. This method is gaining popularity due to the potential for cost savings and efficiency, but it also requires a high level of expertise and communication between all parties.

Choosing the Right Method to Minimize Risks

With the various project delivery methods available, it can be challenging to determine the best one for your construction project. One crucial factor to consider is the level of risk management that each method offers. For example, the traditional design-bid-build method has the potential for the lowest risk, as each phase is completed before moving onto the next. However, the lack of collaboration and communication between parties can increase the risk of errors and delays. On the other hand, methods such as design-build and CMAR have more potential for collaboration and risk management, as the contractor is involved in the project from an earlier stage. This allows for better communication and the ability to identify and mitigate risks before they have a chance to escalate. The DBO method also has the potential for significant risk management, as the contractor is responsible for the project's operation. However, this method requires a high level of expertise and communication between all parties, and the success of the project is highly dependent on the contractor's abilities.When choosing the right method, it is essential to consider the project's size, timeline, budget, and complexity, as well as the level of collaboration and risk management required. It is also crucial to have open and honest discussions with all parties involved to ensure a mutual understanding of roles, responsibilities, and expectations.

Case Studies of Successful Projects Using Various Delivery Methods

To demonstrate the effectiveness of different project delivery methods in risk management, let's take a look at a few case studies. In the first case study, a large infrastructure project was successfully completed using the traditional design-bid-build method. Despite delays and budget overruns, the project was completed within the original allocated budget, highlighting the method's low risk potential. In contrast, a hospital project was completed using the design-build method, resulting in significant cost and time savings due to early contractor involvement and collaboration. The project also experienced a lower number of change orders and design errors, demonstrating the benefits of effective risk management through collaboration. Lastly, a transportation project was completed using the DBO method, resulting in high-quality construction and efficient project management. This method not only involved the contractor in the design and construction phases but also in the operation of the project, allowing for proactive risk management throughout the project's lifespan.These case studies show that effective risk management can be achieved in various project delivery methods, and the key is to choose the method that best suits the project's specific needs and requirements.

In Conclusion

Effective risk management is crucial to the success of any construction project. Therefore, it is essential to carefully consider and compare the different project delivery methods to determine the best fit for your project's risk management needs. As demonstrated through case studies, collaboration, communication, and early involvement of all parties can greatly contribute to mitigating risks and ensuring a successful project outcome.

Chapter 23: Managing Risks in Mega-Projects

Large-scale construction projects, commonly referred to as mega-projects, are becoming increasingly common in today's world. These projects involve unique challenges and often require significant resources, both in terms of time and money. With such high stakes involved, effective risk management is crucial for the successful completion of mega-projects. In this chapter, we will discuss the unique risks involved in mega-projects and explore the challenges that come with managing those risks. We will also provide some solutions and best practices for effectively managing risks in large-scale construction projects.

Unique Risks in Mega-Projects

Mega-projects are not your typical construction projects. They involve massive budgets, complex designs, and involve a large number of stakeholders. These factors bring about unique risks that are not often seen in smaller construction projects. Some of the unique risks in mega-projects include:

- Political and economic risks: Mega-projects are often funded by government entities or involve government contracts. This makes them susceptible to factors such as changes in political climate, funding cuts, and shifting economic conditions. These risks can significantly impact the project budget, timeline, and overall success.

- Design and engineering risks: The complexity of mega-projects often leads to unconventional designs and advanced engineering techniques. While these innovations can result in groundbreaking structures, they also come with their own set of risks. Any flaws in the design or engineering can lead to delays, cost overruns, and even safety hazards.

- Supply chain risks: Mega-projects require a vast amount of resources and materials, which often need to be sourced from various locations. This can make the supply chain vulnerable to delays, shortages, and other risks that can significantly impact the project timeline and budget.

- Technological risks: With the advancements in technology, mega-projects are incorporating more and more technology into their designs and construction methods. While this can result in more efficient and sustainable projects, it also brings about the risk of technological failures, as well as cybersecurity threats.

Challenges and Solutions in Managing Risks in Large-Scale Projects

Managing risks in mega-projects comes with its own set of challenges. These challenges can include:

- Decision making: With so much at stake, decision making in mega-projects can be overwhelming. There are multiple stakeholders involved, each with their own priorities and expectations. This can make it challenging to come to a consensus on risk management decisions.

- Communication: The complexity of mega-projects often means that there are multiple teams and departments involved, each with their own communication channels. This can lead to miscommunication or lack of communication, which can make it difficult to identify and address risks in a timely manner.

- Time and budget constraints: Mega-projects often have strict timelines and budgets, making it challenging to allocate resources to risk management activities. This can result in a lack of preparedness for potential risks, leading to costly delays and budget overruns.So how can these challenges be addressed? Here are some solutions and best practices for managing risks in mega-projects:

- Identify risks early on: The earlier you can identify potential risks, the better prepared you will be to manage them. This is why risk identification should be done in the early stages of the project, with regular revaluation throughout the lifecycle of the project.

- Involve all stakeholders: All stakeholders, including the project team, contractors, and clients, should be involved in the risk management process. This ensures that all perspectives are taken into account, and everyone is working towards a common goal.

- Create a risk management plan: A well-defined risk management plan is crucial for mega-projects. This plan should include risk identification and assessment, as well as strategies for risk mitigation and response. It should also outline roles and responsibilities for each team member involved in risk management.

- Continuous monitoring and evaluation: Risks can change and evolve throughout the project, which is why it is essential to continuously monitor and evaluate risks. This allows for quick identification and response to new risks that may arise.

- Effective communication: Communication is key in any project, but it becomes even more crucial in mega-projects. Establishing clear communication channels and protocols can ensure that important risk-related information is shared and acted upon promptly.

In conclusion, mega-projects bring about unique challenges and risks that require careful attention and proactive management. By identifying risks early, involving all stakeholders, and implementing a well-defined risk management plan, these challenges can be effectively mitigated. By continuously monitoring and evaluating risks and maintaining effective communication, mega-projects can be completed successfully, on time and within budget.

Chapter 24: Identifying and Managing Risks in Public-Private Partnership Projects

In recent years, Public-Private Partnerships (PPP) have become increasingly popular in the construction industry as a way to fund and deliver large infrastructure projects. This unique collaboration between the public and private sectors allows for a more efficient use of resources and expertise, resulting in successful projects that benefit both parties involved. However, with this new model comes new risks, and it is essential for both sides to be aware of these risks and have effective strategies in place to manage them.

Understanding Public-Private Partnerships

Before diving into the risks associated with PPP projects, it is essential to understand the concept of Public-Private Partnerships. Simply put, PPP is a contractual arrangement between a government agency and a private entity to jointly deliver a public service or infrastructure project. This includes a wide range of projects such as transportation, utilities, healthcare facilities, and more. The private partner is responsible for financing, designing, building, and operating the project, while the public partner provides land, regulatory approvals, and long-term payments for the services provided.

Risk Identification in PPP Projects

Like any other form of project delivery, PPP projects also involve a certain level of risk. Identifying and assessing these risks is crucial to the success of the partnership. Some common risks associated with PPP projects include financial risks, legal and regulatory risks, political risks, and construction risks.Financial risks refer to the possibility of inadequate funding, changes in interest rates, and overall financial instability, which can affect the project's cost and timeline. Legal and regulatory risks arise from changes in laws and regulations that can impact the project's operations and obligations. Political risks, on the other hand, stem from a change in government policies or adverse public opinion, which could result in cancellations or delays in the project. Construction risks include unforeseen delays, cost overruns, and design flaws, among others.

Managing Risks in PPP Projects

While there are various risks associated with PPP projects, effective risk management strategies can help mitigate these risks and ensure project success. Here are a few key techniques for managing risks in PPP projects:

1. Establish a Clear Allocation of Risks

One of the first and most crucial steps in managing risks in PPP projects is to establish a clear and fair allocation of risks between the public and private partners. This includes clearly defining who is responsible for each risk and outlining the process for addressing and managing those risks.

2. Conduct Extensive Due Diligence

Before entering into a PPP agreement, both the public and private partners must conduct thorough due diligence to assess the project's feasibility. This includes evaluating the risks involved and ensuring that the project's potential benefits outweigh these risks.

3. Define a Communication Plan

Effective communication is key to managing risks in PPP projects. Both partners should agree on a communication plan that outlines how and when risks will be reported and how communication will be maintained throughout the project's lifecycle.

4. Include Risk Sharing Mechanisms

Including risk-sharing mechanisms in the PPP agreement can help ensure that both partners are equally invested in managing risks. This could include joint contingency funds, performance guarantees, or insurance policies.

Case Studies of Successful Partnerships

While there are undoubtedly risks involved in PPP projects, many successful partnerships have proven that these risks can be effectively managed. Here are two examples of successful PPP projects:

1. Indiana Toll Road

In 2006, the Indiana Toll Road was leased to a private operator for 75 years in exchange for $3.8 billion. This PPP project has been incredibly successful, surpassing its original traffic and revenue projections, resulting in millions of dollars in savings for the state.

2. London Underground

In 2003, Transport for London entered into a 30-year PPP agreement with two private-sector companies to upgrade and maintain the London Underground's infrastructure. This partnership has resulted in significant improvements to the system's efficiency and has been deemed a success by both parties.

Conclusion

Public-Private Partnerships provide a valuable opportunity for both the public and private sectors to work together towards a common goal of delivering crucial infrastructure projects. While risks are inevitable, identifying and managing these risks effectively can lead to successful outcomes for all stakeholders involved. By utilizing the right risk management techniques and learning from successful partnerships, PPP projects can continue to play a significant role in the construction industry's future.

Chapter 25: Effective Communication with Project Teams

Effective communication is vital for the success of any project, especially in the construction industry where there are numerous stakeholders involved. In Chapter 25, we will explore strategies for effective communication, the importance of team collaboration in risk management, and how to manage diverse project teams.

Strategies for Effective Communication

Communication is a key component of risk management in construction projects. Without effective communication, a project can quickly become derailed and face serious consequences. To ensure successful communication, project managers can implement the following strategies:

- Clearly define roles and responsibilities: It is important for every team member to know their role and responsibilities in the project. This will avoid confusion and ensure that everyone knows who to turn to for specific tasks.

- Utilize various communication channels: With the advancement of technology, there are numerous communication channels available such as email, phone, video conferencing, and project management software. Project managers should utilize a combination of these channels to keep all team members informed and updated.

- Establish a communication plan: A structured communication plan can help ensure that information is flowing effectively among team members. The plan should include the frequency, format, and modes of communication.

- Be clear and concise: In a fast-paced construction project, time is of the essence. To avoid misunderstandings or delays, communication should be clear, concise, and to the point.

- Actively listen: Listening is just as important as communicating. Project managers should listen to their team members and encourage open communication to address any concerns or issues that may arise.

Importance of Team Collaboration in Risk Management

Risk management is not a one-person job. It requires collaboration and cooperation among all members of a project team. Here are some reasons why team collaboration is essential for effective risk management:

- Different perspectives: Every team member brings a unique perspective to the table, which can be valuable in identifying and mitigating risks. Collaborating with others can bring new ideas and insights that may have gone unnoticed.

- Shared responsibility: When everyone on the team takes ownership of risk management, it becomes a shared responsibility. This distributed responsibility can help ensure that potential risks are identified and addressed promptly.

- Improved decision making: Collaborating with the team allows for a more comprehensive understanding of risks and their potential impact. This, in turn, leads to better decision-making when it comes to risk mitigation strategies.

- Enhanced communication: Collaboration fosters open communication among team members, which is crucial for addressing risks promptly and effectively.

Managing Diverse Project Teams

In today's globalized world, construction projects often involve a diverse team of individuals with different backgrounds, cultures, and languages. This diversity can bring a wealth of knowledge and skills to a project, but it also presents some challenges in risk management. Here are some tips for managing diverse project teams:

- Recognize and embrace differences: Every team member brings unique skills and experiences to the table, and project managers should embrace this diversity. By recognizing and valuing these differences, team members are more likely to feel included and motivated to contribute to the project's success.

- Establish a common goal: Despite their differences, all team members should have a common goal – the successful completion of the project. By setting this shared objective, team members are more likely to work together towards a common goal.

- Encourage open communication: Diverse team members may have different communication styles, so it is important to encourage open communication and create a safe space for everyone to voice their ideas and concerns.

- Provide cultural sensitivity training: If team members come from different cultures, providing training on cultural sensitivity and understanding can help promote teamwork and reduce misunderstandings.

- Utilize technology: With advancements in technology, communication can be easier and more efficient among diverse team members. Tools such as translation software can help bridge language barriers, while project management software can help streamline communication and collaboration.

Conclusion

Effective communication, team collaboration, and managing diverse teams are crucial components of successful risk management in construction projects. By implementing the strategies discussed in this chapter, project managers can mitigate risks, ensure timely communication, and foster a collaborative and inclusive team environment. When team members work together towards a common goal, they are more likely to overcome challenges and achieve a successful outcome.

Chapter 26: Stakeholder Engagement and Relationship Building for Effective Risk Management in Construction Projects

Effective risk management in construction projects involves not only identifying and assessing risks, but also actively engaging stakeholders and building strong relationships with them. Stakeholders play a critical role in project success, and their engagement can greatly impact the mitigation and resolution of potential risks. In this chapter, we will explore how to conduct a stakeholder analysis, mitigate reputational risks, and build strong stakeholder relationships for effective risk management in construction projects.

Stakeholder Analysis and Engagement

A stakeholder analysis is a crucial step in understanding the interests, concerns, and potential risks associated with each stakeholder involved in a construction project. This analysis provides a comprehensive view of stakeholders and their potential impact on the project, allowing for targeted engagement and risk management strategies. To conduct a stakeholder analysis, first identify all individuals, groups, and organizations that may have a stake in the project. This includes project owners, investors, contractors, subcontractors, suppliers, local communities, government agencies, and more. Once these stakeholders have been identified, categorize them according to their level of influence and interest in the project. High-influence stakeholders are those who have the power to significantly impact the project, while high-interest stakeholders are those who are most affected by the project. Next, assess the potential risks associated with each stakeholder, such as their financial stability, reputation, and past experiences with similar projects. This will help prioritize engagement efforts and identify potential risks that may arise from these stakeholders. It is also important to identify any communication barriers or preferences, as this can affect how effectively the stakeholder can be engaged.Once the stakeholder analysis is complete, it is important to engage stakeholders early and often throughout the project. Encourage open and transparent communication and involve them in decision-making processes. This can help identify and mitigate potential risks early on, before they escalate and impact the project. By engaging stakeholders in a meaningful way, their concerns and interests can be addressed, and their support can be gained, ultimately contributing to the success of the project.

Mitigating Reputational Risks

In the construction industry, reputational risks can have a significant impact on project success. Reputational risks can arise from a wide range of stakeholders, including project owners, contractors, subcontractors, suppliers, and even the local community. These risks can include negative media coverage, public backlash, and damage to a company's overall reputation. To mitigate reputational risks, it is important to first conduct a thorough stakeholder analysis, as mentioned earlier. By understanding the concerns and potential risks of each stakeholder, proactive steps can be taken to address and manage these risks. This includes open and transparent communication, timely and accurate reporting, and actively addressing any issues or concerns raised by stakeholders.Another important strategy is to have a crisis management plan in place. This plan should outline steps to be taken in the event of a crisis or negative event that could impact the project or company's reputation. It should also identify key individuals responsible for managing the crisis and outline communication protocols to ensure a timely and effective response.

Building Strong Stakeholder Relationships

Effective stakeholder engagement and management relies on building strong relationships with stakeholders. This involves understanding their interests, concerns, and expectations, and actively working to address and manage these throughout the project. One way to build strong stakeholder relationships is to establish clear and open lines of communication. This includes having designated individuals responsible for communicating with stakeholders and providing regular updates on project progress. It is also important to actively listen to stakeholder concerns and address them in a timely and transparent manner. Another strategy is to involve stakeholders in decision-making processes. This not only shows them that their opinions and interests are valued, but also allows for their expertise and knowledge to be utilized in the project. By involving stakeholders in decision-making, they will feel more invested in the project and be more likely to support it. In addition, it is important to recognize and appreciate the contributions of stakeholders throughout the project. This can be done through small gestures such as sending thank-you notes or hosting appreciation events. By showing genuine appreciation, stakeholders will feel more connected and committed to the project.Overall, effective risk management in construction projects relies heavily on understanding and engaging stakeholders. By conducting a thorough stakeholder analysis, actively managing reputational risks, and building strong relationships, potential risks can be identified and managed, ultimately contributing to project success.

Chapter 27: Identifying and Managing Risks in Sustainable Construction

Sustainability has become a critical aspect of modern construction projects, with a growing focus on reducing the impact of the built environment on the planet. As a result, it is essential for construction companies to incorporate sustainability into their risk management plans. In this chapter, we will explore the various risks associated with sustainable construction and how to effectively identify and manage them.

Integrating Sustainability into Risk Management Plans

In today's world, sustainable construction practices are no longer just a trend but a necessary step towards a better future. However, these practices come with their own unique set of risks that must be carefully managed. Integrating sustainability into risk management plans not only helps to mitigate these risks but also ensures that projects are completed with minimal impact on the environment.One of the main challenges in integrating sustainability into risk management plans is the lack of understanding and awareness among construction professionals. Often, sustainability is viewed as an added cost rather than an investment in the long-term success of a project. To overcome this challenge, it is essential to educate project teams on the benefits of sustainable practices and how they contribute to risk mitigation.

Identifying Risks in Sustainable Construction

Identifying risks is the first crucial step in any risk management plan, and sustainable construction is no exception. While many risks in sustainable construction may be similar to those in traditional construction, there are also risks unique to sustainable practices. Here are some potential risks to consider during the risk identification process:

- Regulatory compliance risks: With the increasing focus on sustainability, there are often strict regulations and policies in place that must be followed. Failure to comply with these regulations can result in hefty fines and delays in project completion.

- Resource availability risks: Sustainable construction often requires the use of specific materials and resources that may not be readily available in all locations. This can lead to supply chain disruptions and project delays.

- Technology risks: Implementing new technologies and practices to achieve sustainability goals can be risky, as these methods may not have been extensively tested and can lead to unexpected failures or delays.

- Financial risks: Adopting sustainable practices can be expensive, and if not carefully managed, can lead to cost overruns and financial strain on the project.

Managing Risks in Sustainable Construction

Once risks have been identified, it is essential to have a plan in place to effectively manage them. Here are some strategies for managing risks in sustainable construction:

- Partnering with experienced consultants: Sustainable construction practices are constantly evolving, and it is helpful to have experts on the team who can provide guidance and help mitigate risks.

- Conducting thorough research: Before implementing any new sustainable practices or technologies, it is recommended to conduct thorough research and testing to minimize potential failures or setbacks.

- Creating contingency plans: In the event of unexpected supply chain disruptions or technology failures, it is crucial to have contingency plans in place to keep the project on track.

- Regular communication and collaboration: Sustainability can only be achieved through collaboration and effective communication between all project stakeholders. Regular meetings and updates can help identify and mitigate risks in a timely manner.

- Continuous monitoring and evaluation: Sustainable construction is an ongoing process, and risks must be continually monitored and evaluated to ensure they are effectively managed throughout the project's lifecycle.

Incorporating Sustainability into Risk Response Planning

Risk response planning is another critical aspect of risk management in sustainable construction. Here are some ways to incorporate sustainability into risk response planning:

- Risk transfer: Consider partnering with sustainable suppliers and contractors to help mitigate risks associated with resource availability.

- Risk avoidance: Thoroughly research and test new sustainable technologies before implementing them to reduce the risk of failures and delays.

- Risk reduction: Implementing sustainable practices that can reduce waste and improve the project's overall efficiency can help minimize financial risks.

- Risk acceptance: Recognize that sustainability comes with its own set of risks, and have contingency plans in place to manage them effectively.

In Conclusion

Incorporating sustainability into risk management plans is crucial for the success of any construction project in today's world. By identifying and managing risks from the early stages of the project, construction companies can not only ensure the project's success but also contribute towards a more sustainable future. With proper education, awareness, and collaboration, we can mitigate the risks associated with sustainable construction and build a better world for generations to come.

Chapter 28: Identifying and Managing Risks in the Maintenance Phase

The maintenance phase is a crucial aspect of any construction project, as it ensures the long-term functionality and sustainability of a structure. However, it also comes with its own set of risks and challenges that must be carefully managed and addressed. In this chapter, we will delve into the importance of effectively identifying and managing risks during the maintenance phase, implementing preventative maintenance strategies, and examining case studies of successful maintenance risk management.

Identifying Risks in the Maintenance Phase

Identifying potential risks in the maintenance phase is essential for any construction project to ensure its long-term success. These risks can range from structural integrity issues to equipment malfunction to unforeseen environmental factors. It is crucial to have a comprehensive risk assessment at this stage to identify and prioritize potential risks. One effective way to identify risks is to conduct regular inspections and condition assessments of the structure. These assessments can identify any damages or issues that may lead to significant problems in the future. Additionally, involving maintenance staff and contractors in the project planning and design phase can also help in identifying potential risks and developing strategies to mitigate them. Another crucial aspect of risk identification is considering the specific nature of the construction project. For example, a risk assessment for a high-rise building will differ from that of a bridge or a highway. It is essential to have a tailored approach in identifying risks that are specific to the project's type and location.

Managing Risks in the Maintenance Phase

Once potential risks have been identified, it is imperative to have a plan in place to manage and mitigate them. This requires a proactive and comprehensive approach to risk management. Some effective strategies for managing risks in the maintenance phase include:

- Developing a detailed maintenance schedule to ensure regular inspections and repairs are carried out

- Implementing a reporting and documentation system to track any issues that may arise during the maintenance phase

- Investing in proper training and resources for maintenance staff to effectively address any potential hazards

- Communicating effectively with project stakeholders and contractors to ensure prompt action is taken in addressing risks

It is essential to continuously monitor and reassess the risks throughout the maintenance phase and make any necessary adjustments to the risk management plan. Additionally, having a contingency plan in place for potential emergencies or unforeseen events is crucial for successful risk management during the maintenance phase.

Preventative Maintenance Strategies

One of the most effective ways to manage risks during the maintenance phase is by implementing preventative maintenance strategies. These proactive strategies involve conducting regular maintenance and repairs to prevent potential issues from arising in the first place. Some common preventative maintenance strategies include:

- Regularly cleaning and maintaining equipment and machinery

- Conducting routine inspections and repairs of the structure and its components

- Implementing regular maintenance checks for plumbing, electrical, and HVAC systems

- Regularly inspecting and repairing the roof, windows, and doors

- Performing preventive measures against pest infestations

By implementing these preventative maintenance strategies, the risk of potential hazards can be significantly reduced, leading to cost savings and maintaining the structure's overall functionality and sustainability.

Case Studies of Successful Maintenance Risk Management

To further illustrate the importance and effectiveness of effective risk management during the maintenance phase, let's take a look at some real-life case studies of successful implementation:

- The Burj Khalifa, DubaiThe Burj Khalifa, the tallest building in the world, has a comprehensive risk management plan in place for its maintenance. This includes regular inspections and maintenance, replacement of equipment and components before they wear out, and a detailed emergency response plan.

- Golden Gate Bridge, USAThe Golden Gate Bridge has a team dedicated to maintaining and monitoring the bridge's condition 24/7. This team conducts regular inspections and repairs, utilizes advanced technology to identify potential risks, and has a detailed risk management plan in place for emergencies or unforeseen events.

- Sydney Opera House, AustraliaThe Sydney Opera House has a detailed risk management plan in place for its maintenance, including regular inspections, preventative maintenance strategies, and a contingency plan for emergencies. This has helped to ensure the iconic structure's longevity and functionality.

These case studies highlight the importance of effective risk management during the maintenance phase and how it can lead to successful long-term operation of structures.

In conclusion, the maintenance phase is a crucial aspect of any construction project and requires proactive and comprehensive risk management. By consistently identifying potential risks, implementing preventative maintenance strategies, and developing a detailed risk management plan, the longevity and functionality of a structure can be ensured. With proper risk management, the maintenance phase can be a successful and efficient stage of the construction project.

Chapter 29: Effective Risk Management for Small and Medium-Sized Construction Companies

Unique Challenges for Small and Medium-Sized Companies

Managing risk in construction projects is a critical aspect of ensuring successful project delivery. However, it can be a daunting task for small and medium-sized construction companies due to their limited resources and capabilities. These companies often face unique challenges that make risk management even more complex and challenging. Some of these challenges include:

1. Limited Financial Resources

Unlike larger companies, small and medium-sized construction companies may not have the financial resources to invest in sophisticated risk management systems. This can make it difficult for them to implement comprehensive risk management strategies and tools, increasing their vulnerability to risks.

2. Lack of Expertise

Small and medium-sized companies may also lack the expertise and experience in risk management. Many of these companies operate with a small team, and their employees may not have the necessary skills to effectively identify, assess, and manage risks.

3. Limited Access to Information

Small and medium-sized construction companies may not have access to the same level of information and data as larger companies. This could hinder their ability to make informed decisions and adequately anticipate and prepare for potential risks.

Strategies for Minimizing Risks with Limited Resources

Despite the unique challenges faced by small and medium-sized construction companies, it is still possible for them to effectively manage risks and ensure project success. Here are some strategies that these companies can implement to minimize risks with limited resources:

1. Go back to Basics

When it comes to risk management, sometimes the simplest solutions can be the most effective. Small and medium-sized companies can start by reviewing their standard operating procedures and protocols. This will help them identify any gaps or weaknesses that could potentially lead to risks and allow them to take corrective action.

2. Develop a Risk Management Plan

A risk management plan is a crucial tool for identifying, assessing, and managing risks. This plan should outline the company's approach to risk management, including roles and responsibilities, risk evaluation criteria, and response strategies. By having a plan in place, small and medium-sized companies can ensure that all risks are being actively monitored and addressed.

3. Prioritize Risks

With limited resources, it is crucial for small and medium-sized companies to prioritize risks and focus on those with the highest likelihood and impact. This will help them allocate their resources more effectively and address the most critical risks first.

4. Invest in Training and Development

To overcome the lack of expertise in risk management, small and medium-sized companies can invest in training and development for their employees. This will not only enhance their employees' skills but also enable them to take on more responsibilities in risk management.

5. Leverage Technology

Advancements in technology have made risk management more accessible and more efficient. Small and medium-sized companies can leverage various tools and software to identify and monitor risks, such as project management software and risk assessment tools.

6. Collaborate with Larger Companies

Small and medium-sized companies can also collaborate with larger construction companies to gain access to resources and expertise that they may not have. By forming partnerships or joint ventures, these companies can share costs and resources while still maintaining their autonomy.

7. Continuously Evaluate and Improve

Risk management is an ongoing process, and small and medium-sized companies must continuously evaluate and improve their risk management strategies. This includes learning from past projects, identifying areas for improvement, and making necessary adjustments for future projects.

In conclusion, while small and medium-sized construction companies may face unique challenges in risk management, they can still effectively manage risks with the right strategies in place. By prioritizing risks, investing in training and technology, and continuously evaluating and improving, these companies can minimize risks and ensure project success.

Chapter 30: Project Portfolio Risk Management

Managing risk in construction projects is a complex and multifaceted task, requiring careful planning and coordination. However, when companies have multiple projects simultaneously in progress, the risk management process becomes even more challenging. In this chapter, we will delve into the world of project portfolio risk management, exploring strategies for effectively prioritizing risks across multiple projects and sharing lessons learned to ensure the success of future projects.

Project Portfolio Risk Management

Project portfolio risk management involves the management of risks across a company's entire portfolio of projects. This approach allows for a more holistic view of risk, rather than isolating risks to individual projects. By considering potential risks across the entire portfolio, companies can better identify, assess, and mitigate potential risks, ultimately improving overall project performance. One key element of project portfolio risk management is establishing a risk management framework that can be applied to all projects within the portfolio. This includes defining roles and responsibilities, establishing communication protocols, and developing standardized risk assessment and mitigation processes. By having a consistent framework in place, companies can ensure that all projects are following the same risk management standards and protocols. Another critical aspect of project portfolio risk management is regular risk monitoring and reporting. With multiple projects in progress, it can be easy for risks to slip through the cracks. Regular risk monitoring allows companies to quickly identify emerging risks and take proactive measures to mitigate them. It also enables stakeholders to have a holistic view of risk across the entire portfolio and make informed decisions for resource allocation and project prioritization.

Prioritizing Risks Across Multiple Projects

Prioritizing risks across multiple projects requires a strategic and data-driven approach. With limited resources and competing priorities, companies must carefully assess and prioritize which risks to focus on to achieve the best project outcomes. One effective method for prioritizing risks is the use of risk heat maps. These visual representations of risk help stakeholders and project teams quickly identify high-risk areas and allocate resources accordingly. By quantifying and prioritizing risks based on likelihood and impact, companies can make data-driven decisions on how to allocate resources and prioritize project activities. Additionally, project portfolio risk management requires regular communication and collaboration between project teams. It allows for the exchange of information and best practices, enabling teams to learn from each other's experiences and adapt risk management strategies accordingly. This approach can also help identify common risk factors across multiple projects, allowing companies to implement more effective and efficient risk mitigation strategies.

Sharing Lessons Learned Across Projects

Learning from past experiences and mistakes is crucial for continued improvement and success in risk management. Companies can benefit greatly from sharing lessons learned across projects within the portfolio. This not only allows for knowledge transfer between project teams but also helps mitigate risks for future projects. One way to facilitate the sharing of lessons learned is through the use of post-project reviews and evaluations. By conducting thorough assessments of completed projects, companies can identify what worked well and what areas presented challenges. This information can then be shared with other project teams to inform their risk management approach and improve project performance. Furthermore, creating a centralized repository of lessons learned can provide project teams with quick access to valuable information and insights. This can also help identify trends and patterns in risk management across the project portfolio, allowing for a more strategic and proactive approach to risk mitigation.

In conclusion, project portfolio risk management is a critical element for the success of construction projects. By establishing a risk management framework, regularly monitoring and reporting risks, and prioritizing risks across multiple projects, companies can effectively mitigate risks and maximize project outcomes. Additionally, sharing lessons learned across projects can improve risk management strategies and ultimately lead to more successful projects in the future.

Chapter 31: Managing Risks in the Face of Change

The only constant in life is change, and that holds true for construction projects as well. No matter how well-planned and executed a project may be, unexpected changes can occur and disrupt the entire process. These changes can have a significant impact on project risks and can throw the entire project off track. As a project manager, it is essential to be prepared for these changes and have strategies in place to manage them effectively.

Impact of Change on Project Risks

Change can come in many forms during a construction project. It could be changes in scope, design, materials, budget, or even changes in the project team. Each of these changes has the potential to impact project risks in different ways. For instance, a change in design may result in delays and increased costs, while a change in project team members may lead to miscommunications and errors. One of the main impacts of change on project risks is the increase in uncertainty. With any change comes a certain level of uncertainty, and this uncertainty can create new risks or escalate existing ones. As a project manager, it is crucial to identify and assess the potential risks that may arise from any changes and develop a plan to mitigate them.Furthermore, changes can also have a ripple effect on other aspects of the project, causing a domino effect of risks. For example, a change in materials may lead to delays in construction, which could then result in penalties for missing deadlines. This domino effect can have severe consequences for the project if not managed properly.

Strategies for Managing Unexpected Changes

Managing change in a construction project requires a proactive and adaptable approach. Here are some strategies that can help project managers effectively handle unexpected changes and their impact on project risks.

1. Have a flexible project plan:
A well-defined and detailed project plan is essential for any project. However, it is equally important to have some flexibility built into the plan to accommodate unexpected changes. A rigid plan can make it challenging to adapt to changes and may lead to a chain reaction of risks.

2. Communicate openly and frequently:
Good communication is key to managing change effectively. It is crucial to keep stakeholders informed about any changes and how it impacts the project risks. This level of transparency can help build trust and mitigate any potential conflicts that may arise due to changes.

3. Identify potential risks early on:
As soon as a change is identified, it is essential to assess its potential impact on project risks. By identifying risks early on, you can develop mitigation strategies to prevent them from escalating and adversely affecting the project.

4. Utilize risk management tools and techniques:
Risk management tools and techniques, such as risk assessments and contingency planning, can help project managers proactively manage potential risks and their impact on the project. These tools also allow for a more efficient and effective response to changes.

Communicating Changes to Stakeholders

Communication is critical when it comes to managing any changes in a construction project. Stakeholders play a vital role in any project, and their support and understanding are crucial in managing project risks. Here are some communication strategies to consider when communicating changes to stakeholders.

1. Clear and concise communication:
When communicating changes, it is important to be clear and concise. Stick to the facts and provide all the necessary information. This can prevent misunderstandings and help stakeholders make informed decisions.

2. Understand stakeholders' perspective:
Each stakeholder has their own set of priorities and concerns. It is important to

understand their perspective and acknowledge their concerns. This can help build trust and prevent conflicts.

3. Keep stakeholders engaged and involved:
Keeping stakeholders engaged and involved in the project can help them feel invested and aligned with the project's goals. This can also help in managing unexpected changes as stakeholders may be more willing to support and adapt to these changes.

4. Provide regular updates:
Keeping stakeholders informed about the project's progress, including any changes, is crucial for successful project management. Regular updates can also provide stakeholders with a sense of control and understanding of the project's direction.

In conclusion, managing risks in the face of change requires a proactive and adaptable approach. With a well-defined project plan, effective communication strategies, and risk management tools, project managers can successfully navigate unexpected changes and mitigate their impact on project risks. Embracing change and being prepared to handle it can lead to improved outcomes and a more successful project overall.

Chapter 32: Leveraging Data Analytics for Effective Risk Management

As the world becomes increasingly data-driven and technology-driven, the construction industry has seen a rise in the use of data analytics for various purposes. One area where data analytics has proven to be highly effective is in risk management for construction projects. In this chapter, we will explore the benefits of using data analytics in identifying and managing risks, as well as the role of predictive analytics in risk mitigation. We will also discuss the integration of data analysis into risk management processes and the impact it has on project success.

Using Data to Identify and Manage Risks

In the traditional approach to risk management, risks were identified and managed based on the experience and expertise of project managers and team members. While this approach may have worked in the past, it is no longer sufficient in today's complex and rapidly changing construction industry. This is where data analytics comes in. By analyzing large amounts of data from various sources, including project history, weather patterns, material costs, and more, construction professionals can identify potential risks that may not have been evident before. This allows for a more proactive approach to risk management, where preventive measures can be put in place to mitigate the impact of potential risks. Additionally, data can also help identify patterns and trends, allowing for a more accurate assessment of the likelihood and severity of risks.Data analytics also allows for real-time monitoring of risks, providing project teams with up-to-date information on potential threats. This helps project managers make more informed decisions and take timely action to avoid or mitigate risks. By leveraging data analytics, construction companies can better understand the complexities of their projects and make more accurate risk assessments.

Predictive Analytics for Risk Mitigation

One of the most powerful applications of data analytics in risk management is its ability to predict upcoming risks. By analyzing past project data and external factors, such as economic conditions and industry trends, construction professionals can identify potential risks that may arise in the future. This predictive approach to risk management allows project teams to take proactive measures to prevent or reduce the impact of potential risks.Predictive analytics can also help with risk response planning. By analyzing the historical performance of risk mitigation strategies, project teams can determine the most effective and efficient strategies for dealing with specific risks. This not only saves time and resources but also increases the chances of successful risk mitigation.

Integrating Data Analysis into Risk Management Processes

In order to fully capitalize on the benefits of data analytics for risk management, it is crucial to integrate data analysis into the overall risk management processes. This involves incorporating data analysis tools and techniques into risk identification, assessment, and response planning. Data analysis can also help with risk communication by providing visual representations of data, making it easier for stakeholders to understand and prioritize potential risks. This also allows for more effective collaboration among project teams, enabling them to make better decisions based on data-driven insights. Moreover, integrating data analysis into risk management processes allows for continuous improvement. By analyzing data from past projects, construction companies can identify areas for improvement and adjust their risk management strategies accordingly. This ensures project success and can help in creating a culture of continuous improvement within the organization.

In conclusion, data analytics has proven to be a game-changer in the world of risk management for construction projects. By leveraging data, construction professionals can gain valuable insights that enable them to identify and mitigate potential risks, predict future risks, and continuously improve their risk management strategies. As technology continues to advance and data continues to grow, the role of data analytics in risk management will only become more crucial. It is up to construction companies to embrace this technological evolution and use it to their advantage in managing risks and delivering successful projects.

Chapter 33: Managing Difficult Clients and Subcontractors

Dealing with difficult clients and subcontractors is an inevitable part of any construction project. As much as we would like to work with cooperative and understanding individuals, there will always be those who can be challenging to work with. However, as a project manager or construction professional, it is your responsibility to effectively manage these difficult relationships in order to ensure the success of your project. In this chapter, we will explore some tools and techniques for dealing with difficult clients and subcontractors, maintaining professional relationships, and negotiating contracts and agreements.

Tools and Techniques for Dealing with Difficult Clients and Subcontractors

As the saying goes, "prevention is better than cure." The best way to manage difficult clients and subcontractors is to prevent the situation from occurring in the first place. This can be achieved by implementing effective communication and conflict resolution techniques. Regular communication with all parties involved in the project is crucial in order to identify potential issues early on and address them before they escalate. Another effective tool for managing difficult clients and subcontractors is setting clear expectations and boundaries from the beginning. This can be done through written agreements, such as contracts or project charters, that outline the responsibilities and obligations of all parties involved. These formal agreements can serve as a reference point in case of any conflicts or misunderstandings. In cases where conflicts do arise, it is important to handle them calmly and professionally. This can be achieved through active listening, understanding the other party's perspective, and finding a mutually beneficial solution. It may also be helpful to involve a third party mediator or arbitrator to help facilitate the resolution of the conflict.

Maintaining Professional Relationships

Professional relationships play a crucial role in the success of any construction project.

The ability to build strong and positive relationships with clients, subcontractors, and other stakeholders can make all the difference in achieving project objectives. Here are some tips for maintaining professional relationships:

- Communicate effectively and regularly
- Show appreciation for the contributions and efforts of others
- Be respectful and considerate of others' time and priorities
- Be open and transparent in your communication
- Address issues or conflicts promptly and professionally
- Build trust through consistent and reliable behavior

By actively working to maintain professional relationships, you can create a positive working environment that fosters cooperation, collaboration, and ultimately leads to project success.

Negotiating Contracts and Agreements

When it comes to negotiating contracts and agreements with clients and subcontractors, it is important to approach the process with a collaborative mindset. Negotiating should not be seen as a power struggle, but rather a way to find a mutually beneficial solution that meets the needs of all parties involved. During the negotiation process, it is important to clearly communicate your expectations, needs, and concerns. It is also important to actively listen to the other party's perspective and understand their needs and concerns. This can help to identify potential areas of compromise and find a solution that works for everyone. It is also important to carefully review and understand all aspects of the contract before finalizing it. This includes payment terms, deliverables, timelines, and any clauses or provisions that may affect the project. If there are any concerns or discrepancies, it is important to address them before signing the contract to avoid future conflicts.By using effective negotiation techniques and maintaining open communication, you can build strong and mutually beneficial relationships with clients and subcontractors, and ultimately ensure the success of your construction project.

Conclusion

Managing difficult clients and subcontractors is a challenging but necessary aspect of construction projects. By using tools and techniques for effective communication, conflict resolution, and negotiation, and maintaining professional relationships, you can mitigate potential conflicts and ensure the success of your project. Remember, prevention is always better than cure, so it is important to address any issues or conflicts promptly and professionally. By implementing these strategies, you can navigate difficult relationships and achieve project success.

Chapter 34: Managing Conflicting Priorities in Risk Management

Risk management is a crucial aspect of any construction project. It involves identifying potential risks, assessing their impact and likelihood, and implementing strategies to mitigate or eliminate them. However, risk management is not a one-size-fits-all approach, and it must be balanced with the project's objectives. In this chapter, we will explore the importance of balancing project objectives with risk management and how to prioritize risks based on their impact and likelihood.

Balancing Project Objectives with Risk Management

Construction projects are often complex and involve multiple stakeholders, each with their own set of objectives and priorities. At times, these objectives may conflict with each other, making it challenging to balance them with risk management. For instance, the project owner's main objective may be to complete the project within a tight budget and schedule, while the contractor's focus may be on delivering high-quality work. In such a scenario, risk management must strike a balance between meeting project objectives while also mitigating potential risks. One approach to achieving this balance is by involving all stakeholders in the risk management process. This allows for an open and transparent communication channel where all parties can express their concerns and priorities. It also encourages collaboration and cooperation between stakeholders, leading to a more efficient and effective risk management strategy.Another vital aspect of balancing project objectives with risk management is to have a clear understanding of the project's overall goals and objectives. This includes a thorough analysis of the project's scope, budget, and schedule. By having a clear understanding of these crucial project elements, risk management can be tailored to fit the project's specific needs and objectives.

Prioritizing Risks Based on their Impact and Likelihood

Not all risks are created equal, and prioritizing them is essential to the success of a construction project. It is crucial to evaluate and rank risks based on their impact and likelihood, which allows for a more focused and efficient risk management strategy. Here are some key factors to consider when prioritizing risks:

1. Impact: The impact of a risk refers to the negative consequences it may have on the project. This can include delays, budget overruns, safety hazards, and damage to the project's reputation. High-impact risks should be given more attention and resources in the risk management process to minimize their potential impact.

2. Likelihood: The likelihood of a risk refers to the chances of it occurring. Some risks may have a higher probability of happening, while others may be less likely. It is essential to identify and prioritize high-likelihood risks to proactively plan for their mitigation.

3. Interdependency: Some risks may be interconnected, and the occurrence of one may trigger another. These risks must be evaluated and prioritized together to prevent a domino effect.

4. Speed of Impact: Risks that can have an immediate impact on the project should be given a higher priority than those that may have a delayed impact.

5. Resources: Some risks may require more resources, such as time and budget, to mitigate. It is crucial to consider the availability of resources when prioritizing risks. Prioritizing risks allows for a more targeted approach to risk management. By allocating resources and attention to the most critical risks, project managers can effectively manage and mitigate threats to the project's success.In addition to prioritizing risks, it is also crucial to regularly reevaluate and update the risk management plan. As projects evolve, new risks may arise, and existing risks may change in severity. By continuously reviewing and updating the risk management plan, project managers can ensure that potential risks are identified and addressed promptly.

In Conclusion

Balancing project objectives with risk management requires open communication, cooperation, and a clear understanding of the project's goals and objectives. By prioritizing risks based on their impact and likelihood, project managers can effectively manage potential threats and ensure the successful completion of the project. Continuously reviewing and updating the risk management plan is also critical to staying ahead of potential risks and maintaining project success. It is essential to remember that risk management is an ongoing process that requires constant attention and adaptation to ensure a successful and risk-free project.

Chapter 35: Risk Management in Public Sector Projects

In today's world, public sector projects play a pivotal role in shaping our communities and providing essential services to the public. These projects are often complex and have a significant impact on the lives of those they serve. Therefore, it is crucial for these projects to be managed effectively, including anticipating and mitigating potential risks. However, managing risks in public sector projects can pose unique challenges compared to the private sector. In this chapter, we will explore these challenges and discuss strategies for effective risk management in the public sector.

Unique Challenges in Managing Risks in Public Projects

Managing risks in public sector projects presents its own set of challenges that must be carefully navigated to ensure successful project delivery. One of the most significant challenges is the involvement of multiple stakeholders, including government agencies, elected officials, community groups, and the public at large. This diversity of stakeholders brings varying priorities, expectations, and decision-making processes, making it challenging to reach a consensus on risk management strategies. Another challenge in managing risks in public projects is the high level of scrutiny and accountability. Public projects are subject to strict regulations, oversight, and audits, leaving little room for error. Any missteps or failures can result in severe consequences, including legal action and damage to the reputation of those involved. This added pressure can make risk management a delicate and daunting task.Finally, public sector projects often have limited resources and face budget constraints. This can make it challenging to allocate funds for risk management activities, such as risk assessment, mitigation, and contingency planning. As a result, project managers must find creative and cost-effective ways to manage risks while delivering the project within budget.

Strategies for Effective Risk Management in the Public Sector

Despite these challenges, there are several strategies that can help public sector project managers effectively manage risks and deliver successful projects.

1. Stakeholder Engagement and Communication:
As mentioned earlier, involving various stakeholders in public projects can present challenges. However, getting their buy-in and engagement is critical for effective risk management. Project managers should establish clear and open lines of communication with stakeholders, keeping them informed and addressing any concerns they may have. By involving stakeholders in risk assessment and decision-making processes, project managers can harness their knowledge and expertise to identify and mitigate potential risks.

2. Conducting Thorough Risk Assessments:
In public sector projects, the consequences of risks can have far-reaching impacts, both financially and reputationally. Therefore, it is crucial to conduct thorough risk assessments to identify and prioritize potential risks. This process should involve all stakeholders and consider the project's unique characteristics and objectives. Additionally, risk assessments should be updated regularly throughout the project's lifecycle to account for any changes in circumstances.

3. Implementing Effective Risk Mitigation Strategies:
Once risks have been identified and assessed, project managers must develop and implement strategies to mitigate them. This may include developing contingency plans, establishing risk transfer mechanisms, or implementing risk-sharing agreements with contractors. The goal is to reduce the likelihood and impact of risks while staying within the project's budget and timeline.

4. Utilizing Technology and Data Analytics:
Technology and data analytics are powerful tools that can help project managers identify, track, and manage risks. For instance, building information modeling (BIM) technology can aid in identifying potential design issues that may lead to risks. Similarly, data analytics can detect patterns and trends in project data, highlighting potential risks before they materialize.

5. Adopting a Proactive Approach to Risk Management:
In public sector projects, being reactive to risks is not an option. Project managers must take a proactive approach, continually monitoring and assessing risks and adapting strategies as needed. This also involves anticipating and managing political and economic risks that may arise during the project's lifecycle.

In conclusion, managing risks in public sector projects poses unique challenges that require careful consideration and effective strategies. By engaging stakeholders, conducting thorough risk assessments, implementing mitigation strategies, utilizing technology, and adopting a proactive approach, project managers can effectively manage risks and deliver successful projects for the benefit of the public.

Chapter 36: Managing Risks in International Construction Projects

Identifying and Mitigating Risks in International Projects

In today's globalized world, construction companies are increasingly taking on projects in countries outside of their own. While this presents new opportunities for growth and expansion, it also comes with its fair share of risks. Different cultures, laws, and regulations can all pose potential challenges for international construction projects. Therefore, it is crucial for companies to have effective risk management techniques in place to ensure the success of these projects.The first step in managing risks in international projects is to identify them. This involves conducting a thorough risk assessment, taking into consideration factors such as political stability, economic conditions, and cultural differences. One must also consider the project's location and its potential impact on the environment. By identifying potential risks early on, project managers can develop a plan to mitigate or avoid them altogether.

Cultural Differences and Legal Considerations

When working on international projects, it is essential to understand and respect the cultural differences of the host country. These differences can greatly influence the project's success, as they may affect communication, attitudes towards deadlines, and decision-making processes. For example, in some cultures, hierarchy is highly valued, and decisions are made by higher-ups, while in others, decisions are made more democratically. Failing to understand and adapt to these cultural differences can lead to misunderstandings, delays, and ultimately, project failure.Another crucial aspect of managing risks in international projects is ensuring compliance with local regulations. Construction regulations can vary significantly from country to country, and failure to comply with them can result in legal consequences, project delays, and cost overruns. It is important for companies to thoroughly research and understand the laws and regulations of the host country, and to work closely with local authorities to ensure compliance.

Ensuring Compliance with Local Regulations

To ensure compliance with local regulations, companies must stay up-to-date with any changes or updates to laws and regulations in the host country. This can be a challenging task, as laws and regulations are subject to change, and keeping track of these changes can be a time-consuming process. However, failure to comply with local regulations can result in significant delays and financial losses. To streamline this process, companies can invest in specialized software or hire local experts who are well-versed in the regulations of the host country. Another important aspect of compliance is having a solid contract in place with the client and subcontractors. The contract should clearly outline each party's responsibilities, timelines, and any relevant laws or regulations that need to be followed. It is crucial to have a legal advisor review the contract to ensure it is in line with local laws and regulations. In addition to legal compliance, it is also important to adhere to ethical standards in international projects. This includes fair labor practices, environmental sustainability, and transparency in business dealings. Not only does this create a positive reputation for the company, but it also helps to avoid any potential legal or reputational risks.

In conclusion, managing risks in international construction projects requires careful consideration of cultural differences, compliance with local regulations, and ethical practices. By identifying potential risks, understanding and adapting to cultural differences, and staying up-to-date with local laws and regulations, construction companies can mitigate the potential challenges and ensure the successful completion of their international projects.

Chapter 37: Understanding Financial Risks in Construction

Financial risks are an unavoidable aspect of any construction project. They can arise from various factors, such as changes in market conditions, labor and material cost fluctuations, project delays, and unexpected events. These risks can pose a significant threat to the success of a project, leading to cost overruns, delays, and even project failures.However, understanding financial risks in construction and implementing effective strategies to manage them can minimize their impact and ensure the smooth progress of a project. In this chapter, we will explore the different types of financial risks that construction projects may encounter, and how to effectively manage them.

Strategies for Managing Cost-Related Risks

One of the most common financial risks in construction projects is cost-related risks. These risks refer to any potential changes or increases in project costs that may arise during the course of construction. This includes changes in labor or material costs, fluctuations in currency exchange rates, and unexpected events such as natural disasters or political instability. The key to managing cost-related risks is to have a thorough understanding of the project budget and continuously monitor and analyze potential cost changes. This can be achieved through strategies such as creating a detailed budget breakdown, conducting extensive market research, and regularly updating the budget to reflect any changes.Another crucial aspect of managing cost-related risks is effective communication and collaboration among all project stakeholders. This includes contractors, subcontractors, suppliers, and project managers. By maintaining open lines of communication, potential cost issues can be identified and addressed promptly, avoiding any major budget deviations.

Financial Reporting and Budgeting for Risk Management

Financial reporting and budgeting are essential tools for managing financial risks in construction projects. These processes involve tracking and analyzing project costs and comparing them to the initial budget. This enables project managers to identify any deviations and take necessary corrective measures. Regular financial reporting also allows for the early detection of potential cost risks, allowing project managers to address them before they escalate. Additionally, financial reporting can provide valuable insights for future project budgeting and risk management, based on past project performance and cost trends.Budgeting for risk management is another critical aspect of effectively managing financial risks in construction projects. This involves creating a comprehensive and realistic budget with appropriate contingency plans for potential risks. By accounting for potential risks in the initial budget, project managers can minimize the impact of any unforeseen events on the project's financials.

Conclusion

In conclusion, financial risks are a significant aspect of construction projects that require careful management. By understanding the different types of financial risks that may arise and implementing effective strategies to manage them, project managers can ensure the success of their projects. This includes continuously monitoring costs, effective communication and collaboration among stakeholders, and thorough financial reporting and budgeting. By taking a proactive approach to financial risk management, construction projects can achieve their goals within the allocated budget and timeline.

Chapter 38: Legal Disputes and Construction Projects

As with any business venture, construction projects are not immune to legal disputes. In fact, due to the complex nature of construction projects, the potential for disputes is even higher. These disputes can arise between contractors, subcontractors, owners, and other parties involved in the project, and they can result in costly delays, damaged relationships, and even litigation. As risk managers, it is crucial to understand common legal disputes in construction projects and have strategies in place to avoid and resolve them.

Common Legal Disputes in Construction Projects

There are a variety of legal disputes that can arise in construction projects, but some are more common than others. These include:

- Breach of contract: This is the most common type of dispute in construction projects and occurs when one party fails to fulfill their obligations as outlined in the contract.

- Payment disputes: Disagreements over payments can arise between contractors, subcontractors, and other parties involved in the project.

- Scope changes: These disputes arise when there are changes to the project scope, resulting in disagreements over additional costs and delays.

- Construction defects: If a completed project has defects or does not meet the expected quality standards, it can lead to disputes over responsibility and liability.

- Scheduling conflicts: Delays and changes in the project schedule can often lead to disputes between parties.

Strategies for Avoiding and Resolving Disputes

While it may not be possible to completely eliminate disputes in construction projects, there are strategies that can be implemented to minimize their likelihood and effectively resolve them when they do arise. These include:

- Clear and detailed contracts: By drafting comprehensive and clear contracts that outline all parties' responsibilities and obligations, the risk of disputes can be reduced.

- Regular communication: Effective communication among all parties involved in the project is crucial for identifying potential issues early on and finding solutions.

- Thorough documentation: It is important to keep detailed records of all communication, project changes, and payments to have evidence in case a dispute arises.

- Mediation and alternative dispute resolution (ADR): These methods involve bringing in a neutral third party to help facilitate a resolution without going to court.

- Early identification and resolution: Addressing potential issues as soon as they arise can help prevent them from escalating into costly legal disputes down the road.

Working with Legal Counsel

In the event that a legal dispute does occur, it is important to have experienced legal counsel on your side. They can help assess the situation and provide guidance on the best course of action. When selecting legal counsel for construction projects, consider their experience in the industry and their knowledge of the local laws and regulations. It is also important to establish a good working relationship with your legal counsel and involve them in the project from the beginning so they are familiar with all aspects of the project. It is also important to keep in mind that legal disputes can be costly, both in terms of time and money. Therefore, it is crucial to have risk management strategies in place to avoid these disputes as much as possible. This includes having a comprehensive risk management plan that considers potential legal risks and addressing any potential issues before they escalate.

In conclusion, legal disputes are a common occurrence in construction projects, but with proper risk management techniques in place, they can be minimized and effectively resolved. By understanding common disputes, implementing strategies to avoid them, and working with experienced legal counsel, construction projects can be completed successfully and without costly legal battles.

Chapter 39: Risk Management and Continuous Improvement

Incorporating Risk Management into Continuous Improvement Processes

In the fast-paced world of construction, where time is of the essence and budgets are constantly being scrutinized, it is easy for risk management to be seen as a one-time task to be completed before moving on to the next project. However, the most successful construction companies understand the importance of incorporating risk management into their continuous improvement processes. Continuous improvement is a mindset that focuses on constantly evaluating and improving processes, systems, and practices to achieve better results. By incorporating risk management into this process, construction companies can identify potential risks and implement strategies to prevent them from occurring in the future. By integrating risk management into continuous improvement, companies can create a culture of risk awareness and proactivity. This means that everyone, from project managers to workers on the ground, is responsible for identifying and mitigating risks that may arise during the project. One way to incorporate risk management into continuous improvement is to conduct regular risk assessments at different stages of the project. This allows for potential risks to be identified and addressed before they become big issues. In addition, regular communication and collaboration between team members can help identify potential risks and find solutions to mitigate them. Another important element of incorporating risk management into continuous improvement is creating a learning culture within the company. This means that when risks do occur, instead of placing blame, the focus is on understanding what went wrong and finding ways to improve processes for future projects. This not only helps prevent similar risks from occurring in future projects but also promotes a positive and proactive work environment.

Using Lessons Learned to Improve Future Projects

One of the most valuable resources for continuous improvement in risk management is lessons learned from past projects. By analyzing and understanding what went wrong in previous projects, companies can make adjustments and improvements for future projects. It is important for companies to have a system in place for capturing and documenting lessons learned. This can include holding post-project evaluations where team members can provide feedback and share their experiences. By compiling this information, companies can gain insights into the types of risks that have occurred in the past and how they were handled. Analyzing lessons learned can also help identify patterns and recurring risks. This allows for targeted risk management strategies to be implemented in the planning phase of future projects, reducing the likelihood of these risks occurring again. In addition, lessons learned can also highlight successful risk management strategies and best practices that can be incorporated into future projects. This not only helps prevent future risks but also improves overall project outcomes. It is important for companies to have a system in place for sharing lessons learned throughout the organization. By promoting a culture of continuous learning and improvement, companies can ensure that risk management strategies are constantly evolving to meet the changing needs and challenges of the construction industry.

In conclusion, incorporating risk management into continuous improvement processes is essential for the success of construction companies. By creating a culture of risk awareness and proactivity, regularly conducting risk assessments, and utilizing lessons learned, companies can mitigate potential risks and improve project outcomes. This not only benefits the company itself but also promotes a safe and efficient working environment for all involved.

Chapter 40: Risk Management for Infrastructure Projects

Infrastructure projects, such as the construction of roads, airports, and bridges, play a crucial role in a country's economic growth and development. These projects are not only complex and time-consuming, but they also involve a variety of stakeholders and require a significant investment of resources. As a result, the risks associated with infrastructure projects can be substantial and must be effectively managed to ensure project success. In this chapter, we will explore the unique challenges and risks in infrastructure projects and provide strategies for managing them.

Unique Challenges and Risks in Infrastructure Projects

Infrastructure projects are large and involve numerous interdependent activities, making them vulnerable to a wide range of risks. Here are some of the unique challenges and risks that are commonly encountered in infrastructure projects:

Inadequate Planning

Effective planning is crucial for the success of any project, but it is even more critical for infrastructure projects. These projects involve long lead times and have a significant impact on the surrounding environment and communities. Inadequate planning can lead to delays, cost overruns, and negative impacts on the environment and local communities.

Environmental and Regulatory Compliance

Infrastructure projects must comply with a variety of environmental regulations and standards. Failure to comply can result in costly fines, delays, and negative publicity. Moreover, changes in environmental regulations can significantly impact the project's timeline and budget, making it essential to closely monitor and manage these risks.

Weather and Natural Disasters

Infrastructure projects are susceptible to weather and natural disasters, such as floods, hurricanes, and earthquakes. These events can cause significant delays, damage to equipment and materials, and disruption to the project's schedule. It is essential to have contingency plans in place to mitigate these risks and ensure the project stays on track.

Political and Economic Risks

Infrastructure projects are heavily influenced by political and economic factors such as changes in government policies, fluctuation in interest rates, and currency exchange rates. These risks can impact the project budget, timeline, and even its feasibility. Careful risk management and continuous monitoring of these external factors are necessary to minimize the project's exposure to these risks.

Resource Availability

Infrastructure projects require a significant amount of resources, including skilled labor, materials, and equipment. The availability of these resources can fluctuate, leading to delays or increased costs. It is essential to have a robust resource management plan in place, including contingency strategies, to address any potential resource shortages and maintain project progress.

Strategies for Managing Risks in Building and Transport Projects

Risk Assessment and Identification

The first step in managing risk in infrastructure projects is to conduct a thorough risk assessment and identify potential risks. This process involves analyzing the project's scope, identifying potential hazards, and evaluating their likelihood and impact on the project. A comprehensive risk register, updated regularly, is an effective tool for tracking and managing risks throughout the project's lifecycle.

Contingency Planning

Contingency planning is vital for mitigating risks in infrastructure projects. It involves creating a plan to address potential risks and their impact should they occur. This plan should identify actions that can be taken to minimize the impact of risks, such as alternative construction methods, backup resources, or alternative routes. Timely implementation of contingency plans can significantly reduce the project's exposure to risks.

Effective Communication and Stakeholder Engagement

Effective communication and stakeholder engagement are crucial for successfully managing risks in infrastructure projects. All stakeholders, including government agencies, local communities, and project team members, must be kept informed of the project's progress and any potential risks. Additionally, engaging stakeholders in risk management decision-making can improve their buy-in and support for the project.

Continuous Monitoring and Control

Risk management is not a one-time task; it requires continual monitoring and control. Regular project status updates and risk register reviews will allow project managers to identify new risks and proactively address existing ones. Timely response to risks can significantly reduce their impact on the project.

In Conclusion

Infrastructure projects are complex, high-risk undertakings that require careful planning and management. By considering the unique challenges and implementing effective risk management strategies, project managers can minimize risks and ensure successful project outcomes. Continuous monitoring and evaluation throughout the project's lifecycle can help identify and respond to potential risks promptly, maintaining project progress and minimizing financial losses.

Chapter 41: Role of Technology in Streamlining Risk Management Processes

In today's fast-paced and ever-evolving industry, the use of technology has become necessary for effective risk management. With the increasing complexity and size of construction projects, there is a need for advanced tools and techniques to identify, assess, and mitigate risks in a timely and efficient manner.

Integration with Building Information Modeling (BIM)

One of the key developments in the construction industry in recent years has been the adoption of Building Information Modeling (BIM). BIM is a digital representation of a construction project that contains information about its physical and functional characteristics. It allows all stakeholders, including architects, engineers, contractors, and owners, to collaborate and share project information in a centralized integrated platform. BIM not only helps in improving project efficiency and coordination, but it also plays a crucial role in risk management. With BIM, risks can be identified and analyzed at the design stage, allowing for early intervention and mitigation strategies. BIM software can also integrate with risk management tools, enabling real-time risk assessments and updating of project risk registers.Furthermore, BIM can assist in visualizing and simulating potential risks, enabling stakeholders to better understand and address them. This integration of BIM and risk management tools streamlines the risk management process and provides a more holistic approach to risk management in construction projects.

Virtual and Augmented Reality for Risk Management

The use of virtual and augmented reality (VR/AR) technology in construction is gaining popularity, and it has significant potential in risk management. VR/AR technology allows stakeholders to visualize the project in a 3D virtual environment, providing a realistic representation of the project and its potential risks. Through VR/AR simulations, stakeholders can identify potential hazards on the site, assess their severity, and plan mitigation strategies. This technology also enables stakeholders to experience potential risks in a safe and controlled environment, allowing for better preparation and decision-making. Additionally, VR/AR can be used for training and educating construction workers and managers on risk procedures and protocols. This not only improves their understanding of risks but also increases their safety awareness and reduces the likelihood of accidents on site.

Conclusion

In conclusion, the integration of technology in risk management processes has become imperative for the successful delivery of construction projects. BIM and VR/AR tools have revolutionized the way risks are identified, assessed, and managed, providing a more efficient, accurate, and collaborative approach. As technology continues to advance, it is essential for construction companies to embrace it and utilize it in their risk management practices to ensure project success and safety.

Chapter 42: Understanding Political and Economic Risks in Construction Projects

Understanding Political and Economic Factors that Affect Construction Projects

Construction projects are not immune to the influences of politics and economics. In fact, these factors can greatly impact the success and stability of a construction project. Political decisions can have a direct effect on construction projects, such as changes in regulations or policies that may require costly adjustments. Economic factors, such as changes in interest rates or inflation, can also greatly impact the financial aspects of a project. Therefore, it is crucial for construction companies to understand and effectively mitigate these risks in order to ensure the success of their projects.

Strategies for Mitigating Risks due to Changes in Government or Economy

The constantly evolving landscape of politics and economy can pose significant risks to construction projects. However, there are certain strategies that can be implemented to mitigate these risks and ensure the smooth progress of a project.

1. Conduct thorough research and analysis
The first step in mitigating political and economic risks is to conduct a thorough research and analysis of the current political and economic climate. This can involve monitoring government policies and changes in regulations, as well as economic trends and forecasts. By staying informed and updated, construction companies can anticipate potential risks and plan accordingly.

2. Diversify projects and clients
One way to mitigate risks due to changes in government or economy is to diversify projects and clients. This means not relying on one project or one client for a significant portion of the company's revenue. Diversification can help mitigate risks of financial

instability if a project is affected by political or economic changes.

3. Develop strong relationships with government officials
Having a good relationship with government officials can be beneficial in mitigating political risks. By developing a strong rapport with authorities, construction companies can have a better understanding of potential policy changes and may be able to negotiate or find ways to adapt without incurring high costs.

4. Plan for contingencies
When it comes to economic risks, it is important to plan for contingencies. This means having backup plans and alternative strategies in case of unexpected economic changes. By having contingency plans in place, construction companies can mitigate the impact of economic risks on their projects.

5. Stay financially prepared
Another way to mitigate risks due to changes in government or economy is to maintain a strong financial position. This can involve building up a cash reserve to weather any economic downturns, as well as having a solid financial plan in place.

6. Utilize risk management tools
There are various risk management tools and techniques that can help construction companies identify and mitigate potential political and economic risks. This can include risk assessments, stakeholder mapping, and scenario planning. By utilizing these tools, construction companies can better understand and prepare for potential risks.

7. Stay adaptable
In an ever-changing political and economic landscape, it is important for construction companies to stay adaptable. This means being open to change and being able to adjust strategies and plans as needed. By staying adaptable, construction companies can mitigate the risks posed by unexpected political and economic changes.

In conclusion, construction projects are not immune to the effects of politics and economics. However, by understanding these factors and implementing effective risk management strategies, construction companies can mitigate the impact of potential risks and ensure the success of their projects. It is important for construction companies to continuously stay informed and updated in order to anticipate and adapt to potential changes in government and economy. By incorporating these strategies, construction companies can minimize the impact of political and economic risks and achieve successful construction projects.

Chapter 43: Risk Management during Construction Delays

Types and Causes of Construction Delays

Construction delays are a common and often dreaded occurrence in the construction industry. They can be caused by a variety of factors, both internal and external, and can lead to significant financial and schedule setbacks for construction projects. There are several types of construction delays that can occur. These include, but are not limited to:

Weather-Related Delays

Inclement weather, such as heavy rain, snow, or high winds, can cause delays in construction projects. These delays are often unpredictable and can lead to unsafe working conditions, making it difficult for workers to continue construction.

Design and Planning Delays

Delays in the design and planning phase of a project can have a significant impact on construction progress. Poorly designed plans, discrepancies in design drawings, or changes in project scope can all contribute to delays in the construction process.

Material and Equipment Delays

Delays in the delivery of materials and equipment can also cause setbacks in construction projects. This can be due to a variety of factors, such as supplier issues, transportation delays, or incorrect or damaged items being delivered.

Labor Shortages

In recent years, labor shortages have become a major cause of construction delays. As the industry continues to experience a shortage of skilled workers, it has become challenging to find and retain qualified personnel for construction projects.

Unforeseen Site Conditions

At times, unforeseen site conditions can cause delays in construction projects. These can include discovering underground utilities or hazardous materials, unanticipated soil conditions, or encountering unexpected obstacles during the construction process.

Causes of Construction Delays

The causes of construction delays can be numerous and varied, and it is essential to identify and address them early on to minimize their impact on the project. Some common causes of construction delays include:

Poor Communication

Effective communication between all project stakeholders is crucial for a construction project's success. When communication breaks down, crucial decisions and information may be delayed, resulting in project setbacks.

Inadequate Planning and Risk Management

Insufficient planning and risk management can lead to delays in construction projects. When potential risks are not identified and addressed early on, they can quickly escalate into significant issues that can cause project delays.

Poor Project Management

Inadequate project management can also lead to delays in construction projects. From scheduling and budgeting to managing resources and resolving conflicts, effective project management is essential for a project's timely completion.

Scope Changes

Changes in project scope can cause delays as it may require redesigns, additional materials, and a reallocation of resources. It is crucial to have a well-defined scope and a change management process in place to minimize the impact of scope changes on the project timeline.

External Factors

External factors, such as changes in regulations, unforeseen economic conditions, or political instability, can also cause construction delays. These factors are often beyond the control of the project team and can have a significant impact on a project's schedule.

Risk Management Strategies to Minimize Delay Risks

To effectively manage construction delays, it is essential to have a comprehensive risk management plan in place. This plan should address potential risks, and strategies should be in place to minimize the impact of these risks on the project timeline. Some crucial risk management strategies to minimize delay risks include:

Identify and Assess Risks

The first step in effective risk management is to identify and assess potential risks that may impact the project. A thorough risk assessment should be conducted at the beginning of a project and regularly reviewed as it progresses to identify any new risks that may arise.

Develop Contingency Plans

Once risks have been identified and evaluated, contingency plans should be developed to address them. These plans should outline steps to be taken in case a risk materializes and help mitigate its impact on the project schedule.

Regularly Monitor and Control Risks

Risk management is an ongoing process, and risks should be regularly monitored and controlled throughout the project's duration. This involves tracking progress on contingency plans, evaluating emerging risks, and implementing strategies to mitigate or eliminate them.

Maintain Good Communication

Effective communication is essential in risk management. All project stakeholders should be kept informed about potential risks and their impact on the project timeline. This ensures that everyone is on the same page and can work together to minimize delay risks.

Utilize Technology

Advancements in technology have made it easier to identify, monitor, and control risks in construction projects. Utilizing tools such as project management software, drones for site monitoring, or virtual reality for project planning can help mitigate risks and streamline project processes.

Communicating with Stakeholders during Delays

Construction delays can be stressful for all project stakeholders, including clients, contractors, and subcontractors. It is essential to maintain open and transparent communication with them during delays to minimize any potential conflicts and maintain trust and confidence in the project. Some key points to keep in mind when communicating with stakeholders during delays include:

Be Transparent and Honest

Honesty is crucial when communicating about construction delays. Be transparent about the reasons for the delay and provide realistic updates on expected timelines. This will help manage stakeholders' expectations and minimize any potential conflicts.

Explain the Impact of the Delay

It is essential to explain the impact of the delay, both financially and on the overall project timeline. This will help stakeholders understand the situation better and may provide insights into potential solutions to minimize the delay's impact.

Discuss Potential Solutions

When communicating with stakeholders, it is also helpful to discuss potential solutions to minimize the delay. This can include reassessing resources, increasing manpower, or adjusting the project plan to catch up on lost time.

Document Communications

It is crucial to document all communications with stakeholders regarding delays. This will ensure that everyone is on the same page and has a clear understanding of the situation. It can also serve as evidence in case of any disputes or claims related to the delay.Construction delays can be challenging to manage and can have a significant impact on a project's success. However, by implementing effective risk management strategies and maintaining open communication with stakeholders, these delays can be minimized and managed more effectively.

Chapter 44: Identifying and Managing Natural and Man-Made Disaster Risks

Disasters, whether natural or man-made, can strike at any time and have devastating effects on construction projects. From earthquakes and hurricanes to accidents and cyber-attacks, the risks are ever-present and can lead to significant delays, budget overruns, and even total project failure.

Understanding the Risks

The first step in effectively managing disaster risks is to understand and identify them. This involves conducting a thorough risk assessment, considering all potential hazards and their likelihood of occurring. It is essential to involve all stakeholders, from project managers to contractors and subcontractors, in this process to gain a comprehensive understanding of the project's vulnerabilities.Some common natural disaster risks include earthquakes, floods, hurricanes, tornadoes, and wildfires, depending on the project's location. Meanwhile, man-made disasters such as fires, explosions, structural collapses, and cyber-attacks can occur in any project, regardless of its location. By examining the project's geographical location, design, and construction process, it is possible to identify specific risks and plan accordingly.

Strategies for Building Resilient Structures and Processes

Once disaster risks have been identified, the next step is to develop strategies to mitigate and manage them. A critical component of this is to build resilient structures and processes that can withstand potential disasters and quickly recover if impacted. Here are some essential strategies to consider:

Design and Construction Techniques

The design of a project is a crucial factor in its resilience to disasters. Incorporating features such as flexible and robust building materials, proper land use planning, and redundant systems can help mitigate the effects of natural disasters. For example, using reinforced concrete and steel instead of traditional wood-framed structures can make a building more resilient to earthquakes. In addition to the design, the construction process also plays a significant role in building resilience. Employing advanced technologies such as Building Information Modeling (BIM) can help identify potential risks and plan for them in the early stages of construction. Emphasizing proper site preparation, following safety protocols, and conducting regular inspections can also help mitigate disaster risks.

Emergency Response Plan

Having a well-developed and regularly updated emergency response plan is crucial for managing disaster risks. This plan should include procedures for evacuating the site, securing equipment and materials, and communicating with project stakeholders in case of a disaster. It should also outline roles and responsibilities for different team members and establish protocols for contacting emergency services and government authorities.

Backup and Recovery Systems

In the event of a disaster, a backup and recovery system can significantly reduce the impact on a project. This could include offsite data storage, backup generators, and contingency plans for critical project resources. Regular backups and testing of these systems are also essential to ensure their effectiveness during a disaster.

Insurance and Legal Protection

Another crucial strategy for managing disaster risks is to have adequate insurance coverage and legal protection. This can provide financial assistance in the case of a disaster and protect the project from potential lawsuits. It is essential to work closely with insurance and legal professionals to ensure that all potential risks are covered and properly managed.

Cultivating a Culture of Disaster Resilience

In addition to implementing specific strategies, it is vital to cultivate a culture of disaster resilience among all project stakeholders. This involves promoting awareness and education about potential risks and ways to mitigate them. Regular training on disaster response and prevention, along with open communication and collaboration, can go a long way in building a resilient project team.Moreover, it is essential to foster a mindset of adaptability and flexibility in the face of potential disasters. By continuously evaluating and improving risk management processes, the project team can proactively mitigate risks and quickly adapt to any challenges that may arise.

Conclusion

Construction projects are vulnerable to a wide range of disaster risks, both natural and man-made. However, by identifying and understanding these risks and implementing effective strategies, it is possible to build a resilient project that can withstand and recover from potential disasters. By fostering a culture of disaster resilience and continuously evaluating and improving risk management processes, construction professionals can ensure the success of their projects, even in the face of adversity.

Chapter 45: Understanding Cybersecurity Risks in Construction Projects

The construction industry has seen a significant increase in the use of technology and digital tools in recent years. While this has led to increased efficiency and productivity, it has also brought about new risks and challenges, especially in terms of cybersecurity. Construction projects are becoming increasingly vulnerable to cyber attacks, making it crucial for construction companies to understand and manage these risks effectively. In this chapter, we will explore the various aspects of cybersecurity risks in construction projects and discuss strategies for managing them.

Data Protection and Security Measures

One of the biggest cybersecurity risks in construction projects is the theft or compromise of sensitive data. This includes project plans, designs, financial information, and other confidential data. Such breaches can not only lead to financial losses but also damage a company's reputation. Therefore, it is essential to implement robust data protection and security measures to prevent unauthorized access to sensitive information. One necessary security measure is to have strict access controls in place. This means limiting access to confidential data to only authorized personnel. Companies should also consider implementing multi-factor authentication methods, such as biometric scans or one-time passwords, to further enhance security. Another crucial aspect of data protection is to have regular data backups. In case of a cybersecurity attack or data breach, having backup copies of important data can ensure that the project can continue without significant disruptions. These backups should be stored securely, either on the cloud or in physical backups. It is also crucial to have a disaster recovery plan in place to minimize the impact of a cyber attack on the project.Encryption is another critical security measure that plays a vital role in data protection. By encrypting sensitive data, companies can ensure that even if a cyber attack occurs, the data is unreadable and unusable to the attacker. It is essential to use strong encryption algorithms and methods to maintain the integrity of the data.

Strategies for Managing Cybersecurity Risks

While implementing strong security measures is crucial, it is also necessary to have strategies in place to manage cybersecurity risks effectively. Here are some strategies that can help construction companies protect their projects and data from cyber attacks:

1. Conduct Regular Risk Assessments: It is essential to have a thorough understanding of the potential cybersecurity risks to a project. Regular risk assessments can identify vulnerabilities and help companies develop strategies to address them.

2. Create an Incident Response Plan: In the event of a cybersecurity incident, it is critical to have a plan in place to respond quickly and efficiently. This plan should outline the steps to be taken in case of a cyber attack and assign responsibilities to team members.

3. Educate Employees: Employees can be one of the weakest links in a company's cybersecurity defenses. Therefore, it is crucial to educate them about the importance of data protection, password security, and how to spot and report potential cyber threats.

4. Regularly Update Software: Software updates and patches often include security fixes and vulnerabilities. It is essential to keep all software and systems up-to-date to minimize the risk of cyber attacks.

5. Implement Cybersecurity Training: Just as employees need to be educated about cybersecurity, it is also necessary to provide them with specific training on how to recognize and prevent cyber threats. This can include simulated phishing attacks to test their awareness and response.

Understanding Cybersecurity Risks in Construction Projects

As mentioned earlier, the use of technology and digital tools in construction projects has made them more susceptible to cyber attacks. Here are some of the main areas where cybersecurity risks can arise in construction projects:

1. Building Information Modeling (BIM) Systems: BIM systems have become an integral part of modern construction projects, allowing teams to collaborate and share project

data. However, if there are vulnerabilities in the BIM system, it can compromise the entire project.

2. Internet of Things (IoT) Devices: IoT devices, such as sensors and drones, are becoming increasingly common on construction sites. These devices can be vulnerable to cyber attacks if not properly secured, allowing hackers to access project information or even sabotage operations.

3. Cloud Services: Many construction companies are now utilizing cloud services for storage and project collaboration. While convenient, this also means sensitive project data is stored on servers outside of the company's control, making it vulnerable to cyber attacks.

4. Communication Networks: The various communication networks used in construction projects, such as Wi-Fi and radio frequency, can also be susceptible to cyber threats. Hackers can intercept and manipulate communication, leading to data breaches or sabotage of project operations.

In Conclusion

In today's digital age, the risks of cyber attacks on construction projects cannot be ignored. Companies must be proactive in implementing robust data protection and security measures and have strategies in place to manage potential cyber risks effectively. By understanding these risks and taking necessary precautions, construction companies can protect their projects, data, and reputation from cyber threats.

Chapter 46: Identifying and Managing Risks in BIM Projects

Identifying and Managing Risks in BIM Projects

Building Information Modeling (BIM) is a key technology in the construction industry that has revolutionized the way projects are designed, planned, and implemented. BIM allows for a collaborative and integrated approach to project delivery, where all stakeholders can work on a single digital model to improve coordination, reduce errors, and increase efficiency. However, as with any technology, there are inherent risks that come with implementing BIM in a construction project. In this chapter, we will discuss how to identify and manage risks in BIM projects to ensure their successful implementation and use.

Collaborative Risk Management in BIM

One of the main benefits of BIM is the ability for all project stakeholders to collaborate and work on the same digital model. However, this same collaborative aspect can also create potential risks as different parties may have different levels of expertise and understanding of BIM. It is crucial to establish a collaborative risk management process in BIM projects to ensure that everyone is on the same page and working towards minimizing potential risks.One way to facilitate collaborative risk management in BIM is by involving all stakeholders in the early stages of the project. This allows for open communication and discussion of potential risks and how they can be addressed. It is also important to have a clear protocol for decision-making and risk resolution so that any issues can be resolved in a timely and effective manner.

Integrating BIM into Risk Management Processes

BIM should not be seen as a separate entity from traditional risk management processes, but rather as an integral part of it. Just like any other aspect of a construction project, risks associated with BIM should be identified, assessed, and managed using established risk management frameworks. BIM can provide valuable data and insights that can help inform risk assessments and inform risk management decisions. One crucial aspect of integrating BIM into risk management processes is the use of data analytics. BIM can collect a vast amount of data, which can be analyzed to identify potential risks and improve risk management strategies. By utilizing this data, project teams can make more informed decisions and proactively address risks.

Conclusion

In conclusion, while BIM offers numerous benefits in construction projects, it also comes with inherent risks that must be identified and managed. By implementing a collaborative risk management process and integrating BIM into traditional risk management frameworks, construction projects can successfully mitigate potential risks and maximize the benefits of BIM technology. By following these techniques, project teams can ensure the successful implementation of BIM and drive the construction industry towards a more efficient and collaborative future.

Chapter 47: Managing Risks in Design-Build Projects

Design-build projects have become increasingly popular in the construction industry due to their streamlined and integrated approach. In a design-build project, the design and construction phases are combined, allowing for a more collaborative and efficient project delivery. However, this delivery method also presents unique risks that must be carefully managed in order to ensure a successful outcome. In this chapter, we will explore the unique risks of design-build projects, strategies for effectively managing those risks, and case studies of successful design-build projects.

Unique Risks in Design-Build Projects

One of the main risks in design-build projects is the overlap of design and construction. This can create challenges in managing the project timeline and budget, as changes in design can have a direct impact on the construction phase. Additionally, design-build projects require a high level of coordination and communication between the design and construction teams, which can be difficult to achieve if there is not a strong foundation of trust and collaboration. Another risk in design-build projects is the potential for conflicts of interest. In traditional project delivery methods, the owner hires a separate design team and construction team, creating a checks and balances system. However, in a design-build project, the same team is responsible for both design and construction, which can create conflicts of interest and potential for cost-cutting measures that may compromise the quality of the project.Finally, design-build projects also carry the risk of liability. Due to the integrated nature of this project delivery method, it can be challenging to determine who is responsible for errors or delays. This can lead to disputes and potentially costly legal battles.

Strategies for Managing Risks in Integrated Project Delivery

Despite the unique risks involved in design-build projects, there are several strategies that can be implemented to effectively manage these risks. First and foremost, clear and transparent communication is crucial in design-build projects. All parties involved must have a thorough understanding of their roles and responsibilities, as well as the overall project goals. Regular project meetings and open lines of communication can help to mitigate potential conflicts and ensure that everyone is working towards the same objectives. Another key strategy is to establish a strong and trusting relationship between the design and construction teams. This can be achieved through team building exercises, regular collaboration, and clear communication channels. When the team is working together seamlessly, risks are reduced, and the project can run more smoothly.Additionally, it is essential to have a solid risk management plan in place before the project even begins. This plan should identify potential risks, their likelihood, and strategies for mitigating or managing them. Regular risk assessments should also be conducted throughout the project to ensure that risks are being effectively managed.

Case Studies of Successful Design-Build Projects

There have been many successful design-build projects in the construction industry, which have proven the effectiveness of this project delivery method when risks are managed correctly. One notable example is the Dallas/Fort Worth International Airport Terminal D project. This project was delivered using a design-build approach, and it was completed on time and within budget. The design and construction teams worked closely together, communicating openly and proactively managing risks. This collaborative effort resulted in a successful project that received multiple awards for its design and construction.Another successful design-build project is the Guggenheim Museum in Bilbao, Spain. This iconic structure was completed using an integrated project delivery approach, and it has become one of the most visited tourist attractions in the world. The design and construction teams had a strong working relationship, which allowed them to overcome challenges and deliver a stunning and innovative building.

In Conclusion

Design-build projects offer many benefits, including streamlined project delivery and cost savings. However, it is crucial to recognize and effectively manage the unique risks involved in this delivery method. By implementing clear communication, building trusting relationships, and having a solid risk management plan in place, these risks can be mitigated, and design-build projects can be successfully delivered. Through the case studies of successful projects, we can see the importance of these strategies and the potential for a positive outcome when risks are managed effectively.

Chapter 48: Risks Management in Public-Private Infrastructure Partnerships

Infrastructure projects are necessary for the growth and development of societies. However, the demand for new and updated infrastructure often exceeds the financial resources available to governments. This is where Public-Private Partnerships (P3) come into play. P3 projects involve collaboration between the public and private sector to finance, design, build, and operate public infrastructure projects. As with any type of construction project, P3 projects also come with their own set of risks. In this chapter, we will explore the unique risks in P3 projects, the importance of collaborative risk management between public and private partners, and case studies of successful P3 projects.

Unique Risks in P3 Projects

P3 projects present a different set of risks compared to traditional public projects. The involvement of private entities and the complex contractual arrangements in P3 projects create additional risks that must be carefully managed. Some of the unique risks in P3 projects include:

Political Risk

P3 projects are often funded through public resources and are therefore highly dependent on government support. Changes in political leadership or policies can significantly impact the funding and approvals for P3 projects, thereby exposing them to political risk.

Financial Risk

Since P3 projects rely on a mix of public and private funding, there can be financial risks associated with securing financing. This can include delays in securing financing, changes in interest rates, and fluctuations in market conditions.

Legal Risk

The legal framework governing P3 projects is complex and involves multiple parties. This increases the potential for legal risk, including disputes over contract terms, liability, and unforeseen legal issues.

Social Risk

P3 projects can impact communities and local stakeholders in various ways. This can include disruptions to transportation, changes in land use, and community resistance. Failure to manage social risk can lead to delays, increased costs, and damage to the reputation of the project and its stakeholders.

Collaborative Risk Management with Public and Private Partners

Effective risk management is crucial for the success of P3 projects. Given the unique risks in P3 projects, it is imperative that public and private partners work together to identify, assess, and mitigate risks. Collaborative risk management not only improves project outcomes but also helps in building trust and strengthening the relationship between public and private partners.

Early Risk Identification and Allocation

One of the key aspects of risk management in P3 projects is early identification of risks. Public and private partners should work together to identify risks at the earliest stage, even before the contract is signed. This helps in allocating risks to the party best equipped to manage them, thereby reducing the potential for disputes later on.

Risk Allocation through Contracts

P3 projects involve complex contractual arrangements, including revenue-sharing, risk-sharing, and performance guarantees. These contracts should clearly outline the rights and responsibilities of each party in terms of risk management. This helps in avoiding conflicts and delays in project execution.

Regular Communication and Collaboration

Collaborative risk management also involves regular communication and collaboration between public and private partners. This can include joint risk reviews, workshops, and site visits. Regular communication helps in sharing information, addressing concerns, and finding solutions to potential risks.

Shared Risk Management Tools and Techniques

Public and private partners should also consider using shared risk management tools and techniques. For example, using the same risk assessment or project management software can help in aligning risk management efforts and identifying common risks.

Case Studies of Successful P3 Projects

Despite the unique risks in P3 projects, many have been successfully completed, delivering high-quality infrastructure to communities. Let's explore some case studies that demonstrate the effectiveness of collaborative risk management in P3 projects.

Denver International Airport Expansion Project, USA

The expansion of the Denver International Airport was a $500 million project delivered through a P3 agreement between Denver's Department of Aviation and Great Hall Partners. The project involved the replacement of the existing Airport Terminal Building, which had exceeded its capacity, with a new state-of-the-art terminal. The project was successfully completed in 2013, two years ahead of schedule and under budget. Effective collaboration between public and private partners helped in identifying and managing risks throughout the project.

Kempenfelt Water Treatment Plant, Canada

The Kempenfelt Water Treatment Plant was a $200 million project to design, build, and operate a water treatment plant in Ontario, Canada. The project was delivered through a P3 agreement between the City of Barrie and Plenary Group. The project was completed in 2017, and the new plant was able to produce high-quality drinking water for the community. Collaborative risk management played a significant role in the success of this project, with regular communication and sharing of risk management tools.

N1/N2 Ring Roads, Ireland

The N1/N2 Ring Roads project in Ireland was a $250 million project to upgrade a 32-kilometer section of the N1 and N2 national routes. The project was delivered through a P3 agreement between the National Roads Authority and M3 Motorway, a consortium of public and private entities. Despite facing various challenges, the project was completed on time and within budget, thanks to collaborative risk management efforts between public and private partners.

In conclusion, Public-Private Partnerships have emerged as a valuable tool for delivering much-needed infrastructure projects. However, these projects come with their own set of unique risks that require effective risk management. Collaborative risk management between public and private partners is essential for the success of P3 projects and can lead to timely and cost-effective completion of projects. By learning from successful P3 projects and continuously improving risk management practices, we can better ensure the success of future P3 projects.

Chapter 49: Risk Management for Green Building Projects

Green building has become an increasingly popular trend in the construction industry, driven by the demand for sustainable and environmentally friendly structures. As with any construction project, there are inherent risks involved in green building projects that must be identified and managed in order to ensure successful completion and a positive outcome. In this chapter, we will explore the unique risks associated with green building projects and strategies for minimizing these risks.

Identifying and Managing Risks in Sustainable Construction

Green building projects involve an added level of complexity compared to traditional construction projects. This is due to the use of new and innovative technologies, materials and systems that aim to reduce the environmental impact of the building. However, implementing these new practices comes with its own set of risks. One way to identify risks in sustainable construction is to conduct a thorough risk assessment during the planning phase of the project. This involves conducting a detailed analysis of the project, including the design, materials, technologies and methods being used. It is important to involve all stakeholders, including architects, engineers, contractors, and suppliers, in this process to gain a comprehensive understanding of potential risks.Once risks have been identified, they must be managed effectively. This involves developing a risk management plan that outlines how each risk will be addressed and who is responsible for managing it. Regular monitoring and communication among all stakeholders are essential in successfully managing risks in sustainable construction.

Strategies for Minimizing Risks in Green Building Projects

In addition to identifying and managing risks, there are strategies that can be employed to minimize these risks in green building projects. One strategy is to invest in quality materials and technologies. While this may come at a higher cost initially, it can save money in the long run by reducing the likelihood of costly repairs or replacements due to poor quality products. Another strategy is to have a comprehensive contingency plan in place. This includes having backup solutions for potential risks, such as supply chain disruptions, delays, and changes in regulations. It is also important to have a contingency budget to cover unexpected costs that may arise.Collaboration and clear communication among all stakeholders is also crucial for minimizing risks in green building projects. Regular meetings and updates allow for any potential risks to be addressed and resolved before they escalate.

Certification and Verification Risks

One unique risk associated with green building projects is the certification and verification process. In order to be recognized as a green building, projects must go through a certification process, such as LEED (Leadership in Energy and Environmental Design) or BREEAM (Building Research Establishment Environmental Assessment Method). These certifications involve strict guidelines and criteria that must be met in order to be certified. The risk here lies in the possibility of not meeting these criteria and not being certified, or not receiving the desired certification level. This could result in added costs and delays, as well as damage to the reputation of the project and all involved stakeholders.To minimize these risks, it is important to thoroughly research and understand the certification process and requirements beforehand. Engaging with a certified consultant or project manager who has experience with the specific certification can also greatly reduce the risks involved.

In Conclusion

As the demand for sustainable construction continues to rise, it is crucial for project teams to be aware of and effectively manage the risks involved in green building projects. This involves thorough risk identification and planning, investing in quality materials and technologies, and maintaining open communication and collaboration among all stakeholders. By implementing these strategies, green building projects can mitigate potential risks and achieve successful outcomes.

Manufactured by Amazon.ca
Bolton, ON

41291130R00081